T0183868

Lecture Notes in Artificial Intelligence   9546

Subseries of Lecture Notes in Computer Science

More information about this series at http://www.springer.com/series/1244

Martin Atzmueller · Alvin Chin
Frederik Janssen · Immanuel Schweizer
Christoph Trattner (Eds.)

Big Data Analytics in the Social and Ubiquitous Context

5th International Workshop on Modeling Social Media, MSM 2014
5th International Workshop on Mining Ubiquitous and Social Environments, MUSE 2014
and First International Workshop on Machine Learning for Urban Sensor Data, SenseML 2014
Revised Selected Papers

 Springer

Editors
Martin Atzmueller
University of Kassel
Kassel, Hessen
Germany

Alvin Chin
BMW Technology Group
Chicago, IL
USA

Frederik Janssen
Knowledge Engineering
Technische Universität Darmstadt
Darmstadt, Hessen
Germany

Immanuel Schweizer
TU Darmstadt
Darmstadt
Germany

Christoph Trattner
Graz University of Technology
Graz
Austria

ISSN 0302-9743   ISSN 1611-3349   (electronic)
Lecture Notes in Artificial Intelligence
ISBN 978-3-319-29008-9   ISBN 978-3-319-29009-6   (eBook)
DOI 10.1007/978-3-319-29009-6

Library of Congress Control Number: 2015960407

LNCS Sublibrary: SL7 – Artificial Intelligence

Printed on acid-free paper

This Springer imprint is published by SpringerNature
The registered company is Springer International Publishing AG Switzerland

Preface

Ubiquitous and social computing are creating new environments that foster the social interaction of users in several contexts. Concerning social environments, e.g., social media and the social web, there are a variety of social networking environments being implemented in an increasing number of social media applications. For the ubiquitous ones, there are accordingly different small distributed devices and sensors. Overall, ubiquitous and social environments are transcending many diverse domains and contexts, including events and activities in business and personal life, generating massive amounts of data that then require appropriate analysis approaches, methods, and techniques from the area of big data analytics.

This book sets out to explore this area by presenting a number of current approaches and studies addressing selected aspects of this issue. The individual contributions of this book focus on problems related to Big Data analytics in social and ubiquitous contexts, i.e., integrating both ubiquitous data and social media. Methods for the analysis of Big Data in social and ubiquitous contexts concerning both advanced algorithmic solutions as well as analysis techniques focusing on complex data characteristics can then help to advance our understanding of the dynamics and structures inherent to the respective environments.

In these contexts, we present work that tackles issues such as natural language processing in social media, collective intelligence, analysis of social and mobility patterns, and anomaly detection in social and ubiquitous data.

The papers presented in this book are revised and significantly extended versions of papers submitted to three related workshops: The 5th International Workshop on Mining Ubiquitous and Social Environments (MUSE 2014), which was held on September 15, 2014, in conjunction with the European Conference on Machine Learning and Principles and Practice of Knowledge Discovery in Databases (ECML-PKDD 2014) in Nancy, France, the 5th International Workshop on Modeling Social Media (MSM 2014) that was held on April 8, 2014, in conjunction with ACM WWW in Seoul, Korea, and the First International Workshop on Machine Learning for Urban Sensor Data (SenseML 2014). With respect to these three complementing workshop themes, the papers contained in this volume form a starting point for bridging the gap between the social and ubiquitous worlds. Concerning the range of topics, we broadly distinguish between social context, ubiquitous context, and works bridging both.

For the first main theme, we included three works focusing on aspects regarding the social context. We present "Using Wikipedia for Cross-Language Named Entity Recognition" by Eraldo Fernandes, Ulf Brefeld, Roi Blanco, and Jordi Atserias, providing an interesting method for named entity recognition and classification based on exploiting the link structure of Wikipedia. In "On the Predictive Power of Web Intelligence and Social Media" by Nathan Kallus, a novel predictive approach for Web intelligence using Web and social media mining is presented. In another interesting work, "A Latent Space Analysis of Editor Lifecycles in Wikipedia," Xiangju Qin, Derek Greene, and Padraig Cunningham show how topic analysis can be adapted for analyzing editor behavior and for identifying different activity classes.

For the second main theme bridging the social and ubiquitous context, we included two papers. The paper "On Spatial Measures of Geographic Relevance for Geotagged Social Media Content" by Xin Wang, Tristan Gaugel, and Matthias Keller presents spatial measures describing geospatial characteristics of social media content. In "Formation and Temporal Evolution of Social Groups During Coffee Breaks" by Martin Atzmueller, Andreas Ernst, Friedrich Krebs, Christoph Scholz, and Gerd Stumme, the authors describe different group evolution events in social and ubiquitous environments and present an analysis in the scope of academic conferences.

For the third main theme, we included four works targeting ubiquitous contexts. The paper "A Habit Detection Algorithm (HDA) for Discovering Recurrent Patterns in Smart Meter Time Series" by Rachel Cardell-Oliver presents an approach for detecting habits in smart water meter time series. In "RoADS: A Road Pavement Monitoring System for Anomaly Detection Using Smart Phones" by Fatjon Seraj, Berend Jan van der Zwaag, Arta Dilo, Tamara Luarasi, and Paul Havinga, the authors present a method for road pavement analysis resulting in a real-time multiclass road anomaly detector. The paper "Mining Ticketing Logs for Usage Characterization with Nonnegative Matrix Factorization" by Mickaël Poussevin, Nicolas Baskiotis, Vincent Guigue, Patrick Gallinari, and Emeric Tonnelier sets out to analyze activity patterns in an urban transportation network using nonnegative matrix factorization. Finally, in "Context-Aware Location Prediction" by Roni Bar-David and Mark Last, the authors describe an approach for predicting future locations of vehicles utilizing various types of context information.

It is the hope of the editors that this book (a) catches the attention of an audience interested in recent problems and advancements in the fields of big data analytics, social media, and ubiquitous data and (b) helps to spark a conversation on new problems related to the engineering, modeling, mining, and analysis in the field of ubiquitous social media and systems integrating these.

We want to thank the workshop and proceedings reviewers for their careful help in selecting and the authors for improving the submissions. We also thank all the authors for their contributions and the presenters for the interesting talks and the lively discussions at the three workshops. Their efforts and contributions allowed us to produce such a book.

November 2015

Martin Atzmueller
Alvin Chin
Frederik Janssen
Immanuel Schweizer
Christoph Trattner

Organization

Program Committee

Martin Atzmueller	University of Kassel, Germany
Alvin Chin	BMW Group, USA
Bin Guo	Institut Telecom SudParis, France
Sharon Hsiao	Arizona State University, USA
Kris Jack	Mendeley, UK
Frederik Janssen	Knowledge Engineering Group, TU Darmstadt, Germany
Mark Kibanov	University of Kassel, Germany
Harold Liu	Beijing Institute of Technology, China
Eneldo Loza Mencia	Knowledge Engineering Group, TU Darmstadt, Germany
Andreas Schmidt	University of Kassel, Germany
Benedikt Schmidt	TU Darmstadt, Germany
Immanuel Schweizer	TU Darmstadt, Germany
Christoph Trattner	KMI, TU Graz, Austria
Eduardo Veas	TU Graz, Austria
Christian Wirth	TU Darmstadt, Germany
Zhiyong Yu	Fuzhou University, China
Arkaitz Zubiaga	University of Warwick, UK

Contents

Using Wikipedia for Cross-Language Named Entity Recognition

Eraldo R. Fernandes[1], Ulf Brefeld[2]([⊠]), Roi Blanco[3], and Jordi Atserias[4]

[1] Universidade Federal de Mato Grosso Do Sul, Campo Grande, Brazil
[2] Leuphana University of Lüneburg, Lüneburg, Germany
brefeld@leuphana.de
[3] Yahoo! Labs, London, UK
[4] University of the Basque Country, Donostia, Spain

Abstract. Named entity recognition and classification (NERC) is fundamental for natural language processing tasks such as information extraction, question answering, and topic detection. State-of-the-art NERC systems are based on supervised machine learning and hence need to be trained on (manually) annotated corpora. However, annotated corpora hardly exist for non-standard languages and labeling additional data manually is tedious and costly. In this article, we present a novel method to automatically generate (partially) annotated corpora for NERC by exploiting the link structure of Wikipedia. Firstly, Wikipedia entries in the source language are labeled with the NERC tag set. Secondly, Wikipedia language links are exploited to propagate the annotations in the target language. Finally, mentions of the labeled entities in the target language are annotated with the respective tags. The procedure results in a partially annotated corpus that is likely to contain unannotated entities. To learn from such partially annotated data, we devise two simple extensions of hidden Markov models and structural perceptrons. Empirically, we observe that using the automatically generated data leads to more accurate prediction models than off-the-shelf NERC methods. We demonstrate that the novel extensions of HMMs and perceptrons effectively exploit the partially annotated data and outperforms their baseline counterparts in all settings.

1 Introduction

The goal of named entity recognition and classification (NERC) is to detect and classify sequences of strings that represent real-world objects in natural language text. These objects are called *entities* and could for instance be mentions of people, locations, and organizations. Named entity recognition and classification is thus a fundamental component of natural language processing (NLP) pipelines and a mandatory step in many applications that deal with natural language text, including information extraction, question answering, news filtering, and topic detection and tracking [32] and has received a great deal of interest in the past years.

© Springer International Publishing Switzerland 2016
M. Atzmueller et al. (Eds.): MSM, MUSE, SenseML 2014, LNAI 9546, pp. 1–25, 2016.
DOI: 10.1007/978-3-319-29009-6_1

State-of-the-art methods for detecting entities in sentences use machine learning techniques to capture the characteristics of the involved classes of entities. Prominent methods such as conditional random fields [22,23] and structural support vector machines [2,45,47] need therefore to be adapted to annotated data before they can be deployed. Such data is for instance provided by initiatives such as CoNLL[1] that put significant effort in releasing annotated corpora for practical applications in major languages including English, German [40], Spanish, Dutch [39], Italian[2], and Chinese [48]. Although there are corpora for a few minor languages such as Catalan [27], there exist about 6,500 different languages and a large fraction thereof is not covered by NLP resources at all.

From a practitioners point of view, the performance of NERC systems highly depends on the language and the size and quality of the annotated data. If the existing resources are not sufficient for generating a model with the required predictive accuracy the data basis needs to be enlarged. However, compiling a corpus that allows to learn models with state-of-the-art performance is not only financially expensive but also time consuming as it requires manual annotations of the collected sentences. Frequently, the annotation cannot be left to laymen due to the complexity of the domain and it needs to be carried out by trained editors to deal with the pitfalls and ambiguity. In the absence of appropriate resources in the target language, the question rises whether existing corpora in another, perhaps well-studied language could be leveraged to annotate sentences in the target language. In general, cross-lingual scenarios, for instance involving parallel corpora, provide means for propagating linguistic annotations such as part-of-speech tags [12,50], morphological information [43], and semantic roles [35]. In practice, however, creating parallel corpora is costly as, besides annotating the text, sentences need to be aligned so that translation modules can be adapted. Existing parallel corpora are therefore often small and specific in terms of the covered domain.

In this article, we study whether multilingual and freely available resources such as Wikipedia[3] can be used as surrogates to remedy the need for annotated data. Wikipedia, the largest on-line encyclopedia, has already become a widely employed resource for different NLP tasks, including Word Sense Disambiguation [30], semantic relatedness [18] or extracting semantic relationships [38]. So far, only few contributions involving Wikipedia focus on multilingual components such as cross-language question answering [16].

We present a novel approach to automatically generate (partially) annotated corpora for named entity recognition in an arbitrary language covered by Wikipedia. In the remainder, we focus on NERC and note that our approach is directly applicable to other NLP tasks such as part-of-speech tagging and word sense disambiguation. Our method comprises three stages. In the first stage, Wikipedia entries are labeled with the given NERC tag set. The second stage uses Wikipedia language links to map the entries to their peers in the

---

[1] http://ifarm.nl/signll/conll/.

[2] http://evalita.fbk.eu/.

[3] http://www.wikipedia.org/.

target language. The third stage consists of annotating the detected entities in sentences in the target language with their corresponding tag. Note that the methodology leaves entities that are not linked within Wikipedia unannotated. Consequentially, the procedure results in partially labeled corpora which likely contain unannotated entities. We therefore devise two novel machine learning algorithms that are specifically tailored to process and learn from such partially annotated data based on hidden Markov models (HMMs) and structural perceptrons.

Empirically, we demonstrate that with simple extensions, machine learning algorithms are able to deal with this low-quality inexpensive data. We evaluate our approach by automatically generating mono- and cross-lingual corpora that are orders of magnitude larger than existing data sets. Empirically, we observe that using the additional data improves the performance of regular hidden Markov models and perceptrons. The novel semi-supervised algorithms significantly improve the results of their baseline counterparts by effectively exploiting the nature of the partially annotated data.

The remainder is structured as follows. Section 2 reviews related approaches to generate corpora using Wikipedia. We present the automatic generation of cross-lingual corpora using Wikipedia in Sect. 3 and Sect. 4 introduces the machine learning methods for learning from partially labeled data. We report on our empirical results in Sect. 5 and Sect. 6 concludes.

2 Wikipedia-Based Corpus Generation

There are several techniques that classify Wikipedia pages using NERC labels as a preliminary step for different applications. Some of those applications include, for instance, to extend WordNet with named entities [46], to provide additional features for NERC [21], or even to classify Flickr tags [34]. However, how these techniques can be employed to generate tagged corpora is largely understudied. In the remainder of this section, we review three approaches that are related to our work.

Mika *et al.* [31] aim to improve named entity recognition and classification for English Wikipedia entries using key-value pairs of the semi-structured info boxes. Ambiguity is reduced by aggregating observed tags of tokens (the values) with respect to the fields of the info boxes (the keys). Regular expressions are used to re-label the entities. Rather than complete sentences, the final output consists of text snippets around the detected entities. Their approach ignores language links and is therefore restricted to mono-lingual scenarios.

Nothman *et al.* [33] and Richman and Schone [37] propose methods to assign NERC tags to Wikipedia entries by manually defined patterns, key phrases, and other heuristics, respectively. [37] for instance devise key phrases that serve as a simple heuristic for assigning labels to categories and observe reasonable precision by tagging categories containing the words *people from* as *person*, those including the word *company* as *organization*, and those including *country* as *location*, etc. The two approaches focus on extracting completely annotated sentences which results in two major limitations. There is the risk of erroneously annotating tokens due to overly specified rules and heuristics because sentences must be annotated completely and consequentially, large parts of the corpus are discarded because they are likely to contain false negatives (entities which are not annotated). Compared to [37], we take a different approach by only annotating entities with high confidence and leaving the remaining tokens unlabeled. By doing so, our method acts on a much larger data basis. The final models are trained on the partially annotated sentences and render the use of heuristics and manually crafted rules unnecessary.

Alternative approaches to ours are self-training or distant supervision methods. Knowledge bases like Wikipedia are used to automatically extract information (e.g., entities) that are used to train a classifier which is then used to detect even more entities in the resource, etc. [4]. A general problem with self-training is that the initial models are trained on only a few data points and often do not generalize well. As a consequence, erroneous tags enter the training data and may dominate the whole training process.

3   Generating Annotated Corpora Using Wikipedia

This section presents our approach to automatically generate (partially) annotated corpora for named entity recognition and classification. Our method exploits the link structure of Wikipedia as well as the linkage between Wikipedias in different languages. The proposed methodology consists of 3 stages. Firstly, Wikipedia entries in the source language are annotated with the respective NERC tag set (Sect. 3.1). Secondly, the annotated entries are projected into the target language by following cross-lingual Wikipedia links (Sect. 3.2). Thirdly, anchor texts in the target language linking to previously annotated entries are labeled with the corresponding tag (Sect. 3.3).

Figure 1 illustrates the cross-lingual setting for English (left, source language) and Spanish (right, target language). For each language, there are two entries linking to the river *Danube* (Spanish: *Danubio*). The black pointer indicates the language link from the Spanish Wikipedia page to its peer in English. To generate a corpus in Spanish using English as source language we proceed as follows. The mentions of *Danube* are tagged as *location* and propagated by links 1 and 2 to the *Danube* entry. In our simple scenario, the resulting distribution (*person*:0, *location*:2, ...) clearly indicates that this entry should be annotated as a *location*. Using the Wikipedia language link (link 3), the annotation is propagated to the Spanish entry *Danubio* which is also tagged as *location*. Finally, anchor texts in the Spanish Wikipedia of links four and five pointing to *Danubio* are accordingly annotated as locations. We obtain a partial annotation of the Spanish Wikipedia where mentions of *Danubio* are tagged as *location*.

Table 1 shows an exemplary sentence and its partial annotation. According to described procedure, *Danubio* is successfully annotated as a *location*. Words that are not linked to Wikipedia entries such as *que* as well as words that do correspond to Wikipedia entries but have not been processed yet such as *Carlomagno* remain unlabeled. Since not all entries can be resolved, the final corpus is only partially annotated.

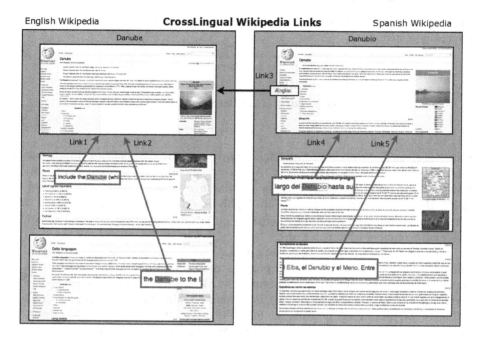

Fig. 1. Wikipedia link structure.

Table 1. A partially annotated sentence.

Carlomagno	contribuyó	a	que	el	*Danubio*	fuese	navegable
?	?	?	?	?	B-LOC	?	?

In the remainder, we focus on the English CoNLL-2003 tags, however we note that the choice of the tags is problem dependent and the incorporation of other tagsets is straight forward. The CoNLL-2003 tags are *PER* (person), *LOC* (location), *ORG* (organization), *MISC* (miscellaneous), and *O* (not an entity). In the following, we introduce our strategy in greater detail.

3.1 Annotating Wikipedia Entries

Our approach to labeling Wikipedia entries with elements of the tag set is based on existing resources. More specifically, we use the freely-available version of Wikipedia from [3], which provides a version of the English Wikipedia with several levels of semantic and syntactic annotations. These annotations have been automatically generated by state-of-the-art algorithms (see [3] for details). Moreover, this corpus preserves the Wikipedia link structure.

Note that the maximal possible number of annotations of the corpus depends on the number of links pointing to the Wikipedia entries in the source language. While some entries such as *Danube* are supported by more than 1,000 mentions

Table 2. Mismatch of annotation (center row) and linked text (bottom row) for an exemplary sentence.

···	are	of	the	Corts		of	Barcelona	from	1283
···	O	O	O	B-MISC	O	B-LOC		O	O
···	O	O	O	wiki/Corts_of_Barcelona			O		wiki/1283

others such as *Holidays with Pay (Agriculture) Convention, 1952* are almost not interlinked at all within Wikipedia. As a consequence, annotations will hardly be generated from this entry and the existing ones may be noisy due to ambiguity. However note that even if only highly interlinked entries are selected, a considerably large set of entries needs to be labeled to build a corpus with a reasonable number of annotations.

We proceed by propagating the annotations to the entries. Every tagged anchor link that is concisely tagged propagates its annotation, while mismatches between linked text and annotation are discarded. Table 2 shows an example of such a mismatch. The link *Cort_of_Barcelona* remains unused because neither *MISC* or *LOC* completely cover the linked text. Recall Fig. 1 for another example. The anchor text linked to the English Wikipedia page *Danube* (links 1 and 2) is tagged as a *location* and the corresponding label *LOC* is assigned to the entry. The major advantages of this approach are twofold. First, it significantly reduces the human effort as no manual annotations are needed. Second, it does not depend on the language dependent category structure. Thus, this approach is generally applicable with any tagged subset of Wikipedia.

Table 3 shows the number of perfect matches between tags and Wikipedia links. Very frequent entries such as *Barcelona* or *Danube* are mostly tagged as locations while others like *Barcelona Olympics* do not show a clearly identifiable peak or are even associated with more than one label as for instance the entry *Barnet*. In general, there are many links to a Wikipedia page and the provided tags are not always perfect. It is thus necessary to reduce the number of ambiguous entities, that is Wikipedia entries that can be associated to more than one tag such as schools which can be either tagged as *organization* or as *location*, depending on the context.

Our approach however naturally allows for detecting ambiguous entities as all occurrences of tags are collected for an entry. Their counts simply serve as indicators to detect ambiguity. We observe a clear peak in the tag-distribution when an entity is not ambiguous; the majority of the annotations correspond to the true class of the respective entities. Thus, a simple strategy to reduce the noise of the tagging and to select a unique label for an entry is to perform a majority voting which corresponds to a *maximum-a-posteriori* prediction given the tag distribution. In practice, it is beneficial to incorporate a threshold θ to select only those Wikipedia entries that have been tagged at least θ-times and to filter entries whose ratio between the first and the second most common label is greater than α.

Table 3. Example of the different counting associated to CoNLL labels

	LOC	PER	ORG	MISC
Danube	1391	31	16	8
Barcelona	3,349	14	1	0
Barcelona Olympics	2	4	2	5
Barnet	33	10	74	0

Table 4 shows the label distribution for $\theta = 30$ and $\alpha = 0.4$ (Wikipedia) for 65,294 Wikipedia entries that are labeled by our method. For comparison, we include the approach by [37] (Category), which maps Wikipedia categories[4] to named entity tags. When applying this tagging technique, we use the same set of key phrases and results from [46], who labeled Wikipedia categories manually. In a few cases multiple assignments are possible; in these cases, we assign the tags to the category matching the most key phrases. Using the category strategy, we obtain NE labels for 755,770 Wikipedia entries. Note that there is no Wikipedia entry assigned to *MISC* as the original list of key phrases does not include a list for the tag *miscellaneous*. Further note that it is generally difficult to define key phrases and detection rules for inhomogeneous entity classes such as *miscellaneous* which are often inalienable in NER as they pool entities that cannot be associated with high confidence to one of the other classes. Another drawback of the category approach is that the entries are found via the Wikipedia category structure and that there is no guarantee for obtaining highly interlinked entries. Recall that the number of annotated entities in our procedure is equivalent to the number of links pointing to entries. The number of resulting annotations cannot be controlled by the category approach. For instance, the category approach leads to 13 M and 800 K entities for the mono-lingual English → English and the cross-language English → Spanish experiments, respectively. While our approach resulted in 19 M and 1.8 M entities, respectively.

3.2   Cross-Lingual Propagation

Once the NERC tags are assigned to Wikipedia entries in the source language, we project these assignments to Wikipedia entries in the target language. This approach exploits the cross-lingual links between Wikipedias in the respective languages, provided that a Wikipedia cross-language link exists between two entries.

Note that links are not bi-directional, that is the existence of a link in one direction does not necessarily imply the existence of the opposite direction. Table 5 shows the number of language links between the English Wikipedia and some smaller Wikipedias in French (1,008,744 entries), Spanish (655,537 entries), Catalan (287,160 entries), Dutch (684,105 entries) and Icelandic (29,727 entries).

---

[4] https://en.wikipedia.org/wiki/Help:Category.

Table 4. Entity distribution for our method (Wikipedia), the category approach [37] (Category), and manually labeled results from [46] (Manual).

	Wikipedia		Category		Manual	
Label	Entries	%	Entries	%	Entries	%
LOC	12,981	19.8	149,333	19,7	404	11.4
ORG	17,722	27.1	107,812	14,2	55	1.5
PER	29,723	45.5	498,625	65,9	236	6.7
MISC	4,868	7.4	-	-	-	-
O	-	-	-	-	2,822	80.2
AMB	-	-	-	-	-	-
	65,294		755,770		3,517	

Table 5. Wikipedia cross-language links.

Links direction	#Links
French → English	730,905
English → French	687,122
Spanish → English	480,336
English → Spanish	475,921
Catalan → English	200,090
English → Catalan	163,849
Dutch → English	167,154
English → Dutch	145,089
Icelandic → English	29,623
English → Icelandic	25,887

Particularly for non-popular languages, the number of cross-lingual links from and to the English Wikipedia varies. Moreover, some of the links are not updated, mistyped, or use different character encodings. For instance, we are only able to map 262,489 Spanish Wikipedia entries out of the 480,336 language links to the corresponding English counterparts. The opposite direction is supported only by 160,918 entries. Nevertheless, apart from the coverage, the cross-lingual propagation can be considered as almost error-free.

3.3 Corpus Annotation

Once the tags are assigned to Wikipedia entries in the target language, the anchor text of the links pointing to tagged entries are annotated with the respective tag. We obtain a partially annotated corpus, as we have no information about annotations for text outside the entity link. The next section deals with machine learning techniques to learn from these partially annotated sentences.

4   Learning from Partially Annotated Data

Traditionally, sequence models such as hidden Markov models [20,36] and variants thereof have been applied to label sequence learning [14] tasks. Learning procedures for generative models adjust the parameters such that the joint likelihood of training observations and label sequences is maximized. By contrast, from an application point of view, the true benefit of a label sequence predictor corresponds to its ability to find the correct label sequence given an observation sequence. Many variants of discriminative sequence models have been explored, including maximum entropy Markov models [29], perceptron re-ranking [2,10,11], conditional random fields [23,24], structural support vector machines [2,47], and max-margin Markov models [45]. In this Section, we present extensions of hidden Markov models and perceptrons that allow for learning from partially labeled data.

A characteristic of the automatically generated data is that it might include *unannotated* entities. For instance, entity mentions may not be linked to the corresponding Wikipedia entry or do not have an associated Wikipedia entry. In cross-language scenarios, linked entries in the target sentences may not be present in the source Wikipedia and thus cannot be resolved. While labeled entities in the automatically generated data are considered ground-truth, the remaining parts of the sentence likely contain erroneous annotations and the respective tokens are thus treated as *unlabeled* rather than *not an entity*.

The following section introduces the problem setting formally. Sections 4.2 and 4.3 present hidden Markov models and perceptrons for learning with partially labeled data, respectively. Section 4.6 discusses ways to parameterize the methods and Sect. 4.7 details their parallelization for distributed computing.

Table 6. An exemplary sentence tagged with the mentioned entities.

	x_1	x_2	x_3	x_4	x_5	x_6	x_7	
$x =$	The	Danube	is	Europe	's	second	longest	$\cdots$
$y =$	0	LOC	0	LOC	0	0	0	$\cdots$
	y_1	y_2	y_3	y_4	y_5	y_6	y_7	

4.1   Preliminaries

The task in label sequence learning [14] is to learn a mapping from a sequential input $x = (x_1, \ldots, x_T)$ to a sequential output $y = (y_1, \ldots, y_T)$, where each observed token $x_t \in \Omega$ is annotated with an element of a fixed output alphabet $y_t \in \Sigma$, see Table 6 for an example. Additionally, we observe some ground-truth annotations of input x denoted by the set $z = \{(t_j, \sigma_j)\}_{j=1}^{m}$ where $1 \le t_j \le T$ and $\sigma_j \in \Sigma$.

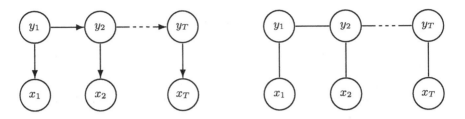

Fig. 2. Hidden Markov model (left) and Markov random field (right) for label sequence learning. The x_t denote observations and the y_i their corresponding latent variables.

Given a sample of n pairs $(\boldsymbol{x}_1, \boldsymbol{z}_1), \ldots, (\boldsymbol{x}_n, \boldsymbol{z}_n)$ the set of labels $\boldsymbol{z}_i$ determine the learning task. If for all $\boldsymbol{z}_i = \emptyset$ holds, observations are unlabeled and the setting is called an *unsupervised* learning task. In case $|\boldsymbol{z}_i| = T_i$ for $1 \leq i \leq n$ all observed tokens are labeled and we recover the standard *supervised* scenario. If sequences are either completely annotated or completely unannotated a *semi-supervised* learning task is obtained, however, the focus of this article lies on learning with *partially annotated* data which generalizes the standard learning tasks and does not make any assumption on the $\boldsymbol{z}_i$. In the remainder we use $\boldsymbol{x}_{[1:t]}$ as a shorthand for the sub-sequence $x_1, \ldots, x_t$ of $\boldsymbol{x}$.

4.2 Hidden Markov Models for Partially Annotated Data

We now extend hidden Markov models (HMMs) to learn from partially annotated data. The novel method combines supervised and unsupervised learning techniques for HMMs and we briefly review HMMs and the Baum-Welch algorithm in Sect. 4.2, respectively.

Hidden Markov Models. Hidden Markov models are generative sequential models [36]. Their underlying graphical model describes how pairs $(\boldsymbol{x}, \boldsymbol{y})$ are generated and is depicted in Fig. 2 (left). That is, a (first-order) hidden Markov model places an independence assumption on non-adjacent factors and computes the joint probability $P(\boldsymbol{x}, \boldsymbol{y})$ by

$$P(\boldsymbol{x}, \boldsymbol{y}) = P(y_1) \prod_{t=1}^{T} P(x_t|y_t) \prod_{t=1}^{T-1} P(y_{t+1}|y_t).$$

Priors $\pi_\sigma = P(y_1 = \sigma)$, transition probabilities $A = (a_{\sigma\tau})_{\sigma,\tau \in \Sigma}$ with $a_{\sigma\tau} = P(y_{t+1} = \tau|y_t = \sigma)$ and observation probabilities $B(\boldsymbol{x}) = (b_\sigma(x_t))_{\sigma \in \Sigma, 1 \leq t \leq T}$ with $b_\sigma(x_t) = P(x_t|y_t = \sigma)$ need to be adapted to the data. Usually, the parameters $\theta = (\pi, A, B)$ are estimated by maximizing the log-likelihood

$$\theta^* = \underset{\theta}{\operatorname{argmax}} \sum_{i=1}^{n} \log P(\boldsymbol{x}, \boldsymbol{y}|\theta).$$

Once optimal parameters θ^* have been found, they are used as plug-in estimates to compute label distributions for unannotated sequences by means of the Forward-Backward algorithm [36]. This algorithm consists of a left-to-right pass computing $\alpha_t(\sigma)$ and a right-to-left pass that computes $\beta_t(\sigma)$. The auxiliary variables are defined as

$$\alpha_t(\sigma) = P(\boldsymbol{x}_{[1:t]}, y_t = \sigma | \theta) = \begin{cases} \pi_\sigma b_\sigma(x_1) & : \quad t = 1 \\ \sum_\tau [\alpha_t(\tau) a_{\tau\sigma}] b_\sigma(x_{t+1}) & : \quad \text{otherwise} \end{cases}$$

$$\beta_t(\sigma) = P(\boldsymbol{x}_{[t+1:T]} | y_t = \sigma, \theta) = \begin{cases} 1 & : \quad t = T \\ \sum_\tau a_{\sigma\tau} b_\tau(x_{t+1}) \beta_{t+1}(\tau) & : \quad \text{otherwise,} \end{cases}$$

and the probability for y_t taking label σ is given by

$$P(y_t = \sigma | \boldsymbol{x}, \theta) = \frac{\alpha_t(\sigma)\beta_t(\sigma)}{\sum_\tau \alpha_t(\tau)\beta_t(\tau)}.$$

Expectation Maximization. In the absence of (partial) labels, that is $\bigcup \boldsymbol{z} = \emptyset$, only unlabeled input sequences $\boldsymbol{x}_1, \ldots, \boldsymbol{x}_n$ are given. In this unsupervised case the Baum-Welch algorithm [5] is often used to learn parameters of hidden Markov models. The algorithm takes the number of possible states as input parameter and initializes the first model randomly. It then maximizes the data likelihood by an Expectation-Maximization (EM) procedure [13] consisting of two alternating steps. The *Expectation*-step computes the most likely annotations for the unlabeled sequences given the input sequences and the model. The *Maximization*-step re-estimates the model parameters given the input sequences and the previously computed annotations. The method is a variant of self-training and converges to a local optimum.

Hidden Markov Models for Partially Annotated Data. We now propose an extension of hidden Markov models that learns from partially labeled data $\{(\boldsymbol{x}_i, \boldsymbol{z}_i\}_{i=1}^n$. The distribution of the labels $\boldsymbol{z}_i$ can be arbitrary and if $|\boldsymbol{z}_i| = T_i$ for all i or in case $\bigcup \boldsymbol{z}_i = \emptyset$ we recover the supervised and unsupervised hidden Markov models as special cases, respectively.

The idea is to revise the Expectation-Maximization framework as follows and grounds on the observation that annotated tokens do not need to be estimated during the *Expectation*-step. Conversely, we may use original EM updates for treating unannotated tokens as these may need to be re-estimated. This observation can be incorporated into the Forward-Backward procedure by altering the definition of the involved probabilities α and β so that the modified variants always chooses the ground-truth label for annotated tokens. The Maximization-step is identical to the original Baum-Welch algorithm but uses the modified $\tilde{\alpha}$ and $\tilde{\beta}$ variables. The modified variables are defined as

$$\tilde{\alpha}_t(\sigma) = P(\boldsymbol{x}_{[1:t]}, \boldsymbol{z}_{\leq t}, y_t = \sigma | \theta) = \begin{cases} 0 & : \quad \text{if } (t, \tau) \in \boldsymbol{z} \wedge \tau \neq \sigma \\ \alpha_t(\sigma) & : \quad \text{otherwise} \end{cases}$$

$$\tilde{\beta}_t(\sigma) = P(\boldsymbol{x}_{[t+1:T]}, \boldsymbol{z}_{>t} | y_t = \sigma, \theta) = \begin{cases} 0 & : \quad \text{if } (t, \tau) \in \boldsymbol{z}(y_t) \wedge \tau \neq \sigma \\ \beta_t(\sigma) & : \quad \text{otherwise.} \end{cases}$$

where $z_{\leq t} = \{(t', \tau) \in z : t' \leq t\}$ denotes the set of annotated tokens up to position t and $z_{>t} = z \setminus z_{\geq t}$ are the labeled tokens at positions greater than t. Marginalizing over the unannotated positions gives us the desired quantities; the distribution of labels at position t is for instance given by

$$P(y_t = \sigma | \boldsymbol{x}, \boldsymbol{z}, \theta) = \frac{\tilde{\alpha}_t(\sigma) \tilde{\beta}_t(\sigma)}{\sum_\tau \tilde{\alpha}_t(\tau) \tilde{\beta}_t(\tau)}.$$

The above computation schema enforces $P(y_t = \sigma | \boldsymbol{x}, \boldsymbol{z})$ for every annotated token $(t, \sigma) \in \boldsymbol{z}$ and $P(y_t = \tau | \boldsymbol{x}, \boldsymbol{z}) = 0$ for alternative labels $\tau \neq \sigma$. For unlabeled tokens x_t the original Expectation-Maximization updates are used. Note that this algorithm is a special case of [41].

4.3   Structured Perceptrons for Partially Labeled Data

The sequential learning task can alternatively be modeled in a natural way by an undirected Markov random field where we have edges between neighboring labels and between label-observation pairs, see Fig. 2 (right). The conditional density $P(\boldsymbol{y}|\boldsymbol{x})$ factorizes across the cliques [19] and different feature maps can be assigned to the different types of cliques, that is ϕ_{trans} for transitions and ϕ_{obs} for emissions [2,23]. Finally, interdependencies between $\boldsymbol{x}$ and $\boldsymbol{y}$ are captured by an aggregated joint feature map $\phi : \mathcal{X} \times \mathcal{Y} \to \mathbb{R}^d$,

$$\phi(\boldsymbol{x}, \boldsymbol{y}) = \left(\sum_{t=2}^{T} \phi_{trans}(\boldsymbol{x}, y_{t-1}, y_t)^\top, \sum_{t=1}^{T} \phi_{obs}(\boldsymbol{x}, y_t)^\top \right)^\top.$$

We are only interested in the maximum-a-posteriori label-sequence which gives rise to log-linear models of the form

$$P(\boldsymbol{y}|\boldsymbol{x}) \propto \boldsymbol{w}^\top \phi(\boldsymbol{x}, \boldsymbol{y}).$$

The feature map exhibits a first-order Markov property and as a result, decoding can be performed by a Viterbi algorithm [17,42] in $\mathcal{O}(T|\Sigma|^2)$,

$$\hat{\boldsymbol{y}} = f(\boldsymbol{x}; \boldsymbol{w}) = \operatorname*{argmax}_{\tilde{\boldsymbol{y}} \in \mathcal{Y}(\boldsymbol{x})} \boldsymbol{w}^\top \phi(\boldsymbol{x}, \tilde{\boldsymbol{y}}). \tag{1}$$

In the remainder, we will focus on the 0/1- and the Hamming loss to compute the quality of predictions,

$$\ell_{0/1}(\boldsymbol{y}, \hat{\boldsymbol{y}}) = 1_{[\boldsymbol{y} \neq \hat{\boldsymbol{y}}]}; \qquad \ell_h(\boldsymbol{y}, \hat{\boldsymbol{y}}) = \sum_{t=1}^{|\boldsymbol{y}|} 1_{[y_t \neq \hat{y}_t]} \tag{2}$$

where the indicator function $1_{[u]} = 1$ if u is true and 0 otherwise.

4.4   Loss-Augmented Structured Perceptrons

The structured perceptron [2,10] is analogous to its univariate counterpart, however, its major drawback is the minimization of the 0/1-loss which is generally
too coarse for differentiating a single mislabeled token from completely erroneous annotations. To incorporate task-dependent loss functions into the learning process, we make use of the structured hinge loss of a margin-rescaled SVM
[28,47].

Given a sequence of fully labeled $(x_1, z_1), (x_2, z_2), \ldots$ where $|z_i| = T_i$, the
structured perceptron generates a sequence of models $w_0 = 0, w_1, w_2, \ldots$ as
follows. At time t, the loss-augmented prediction is computed by

$$\hat{y}_t = \underset{\tilde{y} \in \mathcal{Y}(x_t)}{\operatorname{argmax}} \left[\ell(y_t, \tilde{y}) - w_t^\top \phi(x_t, y_t) + w_t^\top \phi(x_t, \tilde{y}) \right]$$

$$= \underset{\tilde{y} \in \mathcal{Y}(x_t)}{\operatorname{argmax}} \left[\ell(y_t, \tilde{y}) + w_t^\top \phi(x_t, \tilde{y}) \right].$$

Rescaling the margin with the actual loss $\ell(y_t, \tilde{y})$ can be intuitively motivated
by recalling that the size of the margin $\gamma = \tilde{\gamma}/\|w\|$ quantifies the confidence in
rejecting an erroneously decoded output $\tilde{y}$. Re-weighting $\tilde{\gamma}$ with the current
loss $\ell(y, \tilde{y})$ leads to a weaker rejection confidence when y and $\tilde{y}$ are similar,
while large deviations from the true annotation imply a large rejection threshold.
Rescaling the margin by the loss implements the intuition that the confidence
of rejecting a mistaken output is proportional to its error.

An update is performed if the loss-augmented prediction $\hat{y}_t$ does not coincide
with the true output y_t; the update rule is identical to that of the structured
perceptron and given by

$$w_{t+1} \leftarrow w_t + \phi(x_t, y_t) - \phi(x_t, \hat{y}_t).$$

After an update, the model favors y_t over $\hat{y}_t$ for the input x_t, however, note
that in case $\hat{y}_t = y_t$ the model is not changed because $\phi(x_t, y_t) - \phi(x_t, \hat{y}_t) = 0$
and thus $w_{t+1} \leftarrow w_t$.

Margin-rescaling can always be integrated into the decoding algorithm when
the loss function decomposes over the latent variables of the output structure as
it is the case for the Hamming loss in Eq. (2). After the learning process, the
final model w^* is a minimizer of a convex-relaxation of the theoretical loss (the
generalization error) and given by

$$w^* = \underset{w}{\operatorname{argmin}} \, \mathbb{E} \left[\max_{\tilde{y} \in \mathcal{Y}(x)} \ell(y, \tilde{y}) - w^\top (\phi(x, y) - \phi(x, \tilde{y})) \right].$$

4.5   Transductive Perceptrons for Partially Labeled Data

We derive a straight-forward transductive extension of the loss-augmented perceptron that allows for dealing with partially annotated sequences and arbitrary
(partial) labelings z [15]. The idea is to replace the missing ground-truth with

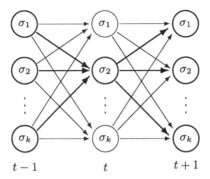

Fig. 3. The constrained Viterbi decoding (emissions are not shown). If time t is anno-tated with σ_2, the light edges are removed before decoding to guarantee that the optimal path passes through σ_2.

a pseudo-reference labeling for incompletely annotated observation sequences. We thus propagate the fragmentary annotations to unlabeled tokens so that we obtain the desired reference labeling as a makeshift for the missing ground-truth. Following the transductive principle, we use a constrained Viterbi algorithm [7] to decode a pseudo ground-truth $\boldsymbol{y}_p$ for the tuple $(\boldsymbol{x}, \boldsymbol{z})$,

$$\boldsymbol{y}_p = \operatorname*{argmax}_{\tilde{\boldsymbol{y}} \in \mathcal{Y}(\boldsymbol{x})} \boldsymbol{w}^\top \phi(\boldsymbol{x}, \tilde{\boldsymbol{y}}) \quad \text{s.t.} \quad \forall (t, \sigma) \in \boldsymbol{z}: \quad \tilde{y}_t = \sigma.$$

The constrained Viterbi decoding guarantees that the optimal path passes through the already known labels by removing unwanted edges, see Fig. 3. Assuming that a labeled token is at position $1 < t < T$, the number of removed edges is precisely $2(k^2 - (k-1)k)$, where $k = |\Sigma|$. Algorithmically, the con-strained decoding splits sequences at each labeled token in two halves which are then treated independently of each other in the decoding process.

Given the pseudo labeling $\boldsymbol{y}_p$ for an observation $\boldsymbol{x}$, the update rule of the loss-augmented perceptron can be used to complement the transductive perceptron. Note that augmenting the loss function into the computation of the argmax gives $\boldsymbol{y}_p = \hat{\boldsymbol{y}}$ if and only if the prediction $\hat{\boldsymbol{y}}$ fulfills the implicit loss-rescaled margin criterion and $\phi(\boldsymbol{x}, \boldsymbol{y}_p) - \phi(\boldsymbol{x}, \hat{\boldsymbol{y}}) = \boldsymbol{0}$ holds.

Analogously to the regular perceptron algorithm, the proposed transductive generalization can easily be kernelized. Note that the weight vector at time t is given by

$$\boldsymbol{w}_t = \boldsymbol{0} + \sum_{j=1}^{t-1} \phi(\boldsymbol{x}_j, \boldsymbol{y}_j^p) - \phi(\boldsymbol{x}_j, \tilde{\boldsymbol{y}}_j)$$

$$= \sum_{(\boldsymbol{x}, \boldsymbol{y}_p, \hat{\boldsymbol{y}})} \alpha_{\boldsymbol{x}}(\boldsymbol{y}_p, \hat{\boldsymbol{y}}) \left[\phi(\boldsymbol{x}, \boldsymbol{y}_p) - \phi(\boldsymbol{x}, \hat{\boldsymbol{y}}) \right] \tag{3}$$

with appropriately chosen α's that act as virtual counters, detailing how many times the prediction $\hat{\boldsymbol{y}}$ has been decoded instead of the pseudo-output $\boldsymbol{y}_p$ for an observation x. Thus, the dual perceptron has virtually exponentially many parameters, however, these are initialized with $\alpha_x(y, y') = 0$ for all x, y, y' so that the counters only need to be instantiated once the respective triplet is actually seen. Using Eq. (3), the decision function depends only on inner products of joint feature representations which can then be replaced by appropriate kernel functions $k(x, y, x', y') = \phi(x, y)^\top \phi(x', y')$.

4.6 Parametrization

The presented extensions of hidden Markov models and structural perceptrons learn from labeled and unlabeled tokens. In practical applications, the unlabeled tokens usually outnumber the labeled ones and thus dominate the optimization problems and consequentially valuable label information does only have little or no impact at all on the final model. A remedy is to differently weight the influence of labeled and unlabeled data or to increase the influence of unlabeled examples during the learning process.

For the hidden Markov models, we introduce a mixing-parameter $0 \leq \lambda \leq 1$ to balance the contribution of labeled and unlabeled tokens such that the final model can be written as

$$HMM_{final}(\mathcal{D}) = (1 - \lambda)HMM_S(\mathcal{D}_\mathcal{L}) + \lambda HMM_U(\mathcal{D}_U),$$

where $HMM_S(\mathcal{D}_L)$ and $HMM_U(\mathcal{D}_U)$ correspond to supervised (HMM_S) and unsupervised (HMM_U) HMMs which are solely trained on the labeled part $\mathcal{D}_L$ and unlabeled part $\mathcal{D}_U$ of the data $\mathcal{D} = \mathcal{D}_L \bigcup \mathcal{D}_U$, respectively. For the perceptrons, we parameterize the Hamming loss to account for labeled and unlabeled tokens,

$$\ell_h(y_p, \hat{y}) = \sum_{t=1}^{|y_p|} \lambda(z, t) 1_{[y_t^p \neq \hat{y}_t]}$$

where $\lambda(z, t) = \lambda_L$ if t is a labeled time slice, that is $(t, \cdot) \in z$, and $\lambda(z, t) = \lambda_U$ otherwise. Appropriate values of λ_{HMM}, λ_L and λ_U can be found using cross-validation or using holdout data.

4.7 Distributed Model Generation

The discussed hidden Markov models and perceptrons can easily be distributed on several machines. For instance, EM-like algorithms process training instances one after another and store tables with counts for each instance in the Estimation-step. The counting can be performed on several machines in parallel as the tables can easily be merged in a single process before the Maximization-step which is again a single process. After the optimization, the actual model is distributed across the grid for the next Expectation-step.

Perceptron-like algorithms can be distributed by using the results by Zinkevich et al. [52]. The idea is similar to that of EM-like algorithms.

Table 7. Descriptive statistics for the English $\rightarrow$ English corpora.

	CoNLL	Wikipedia
Tokens	203,621	1,205,137,774
Examples	14,041	57,113,370
Tokens per example	14.5	21.10
Entities	23,499	19,364,728
Entities per example	1.67	0.33
Examples with entity	79.28 %	21.28 %
MISC entities	14.63 %	13.78 %
PER entities	28.08 %	29.95 %
ORG entities	26.89 %	32.80 %
LOC entities	30.38 %	23.47 %

Equation (3) shows that the order of the training examples is not important as long as counters store the number of times they have been used for updates. Thus, the model generation can be distributed across machines and a final merging process computes the joint model which is then distributed across the grid for the next iteration.

4.8   Related Work

Learning with partially labeled data generalizes semi-supervised learning [8] which aims at reducing the need for large annotated corpora by incorporating unlabeled examples in the optimization. Semi-supervised structural prediction models have been proposed in the literature by means of Laplacian priors [1,24], entropy-based criteria [25], transduction [51], co-training [6], self-training [26], or SDP relaxations [49]. Although these methods have been shown to improve over the performance of purely supervised structured baselines, they do not reduce the amount of required labeled examples significantly as it is sometimes the case for univariate semi-supervised learning. One of the key reasons is the variety and number of possible annotations for the same observation sequence; there are $|\Sigma|^T$ different annotations of a sequence of length T with tag set Σ and many of them are similar to each other in the sense that they differ only in a few labels. Furthermore, the above mentioned methods hardly scale for Wikipedia-sized data sets. Thus the closest method to the proposed extension of the structural perceptron is [44]. Both approaches rely on the same underlying graphical model and types of features, and use EM-like optimization strategies. We thus consider them as of the same family of approaches and note that our approach is conceptionally simpler than the one presented in [44].

5   Evaluation

In this section we report on our empirical evaluation of the automatic corpus generation. The remainder of this section is organized as follows. Section 5.1 summarizes the CoNLL data sets and Sect. 5.2 details our experimental setup. We report on mono-lingual results for English in Sect. 5.3 and summarize the cross-lingual experiments in Sect. 5.4.

5.1   CoNLL Corpora

We use the English, Spanish and Dutch versions of Wikipedia to evaluate our system since manually annotated corpora are available for these languages. We use the corpora provided by CoNLL shared tasks in 2002 [39] and 2003 [40]. The CoNLL'2003 shared task [40] corpus for English includes annotations of four types of entities: person (PER), organization (ORG), location (LOC), and miscellaneous (MISC). This corpus is assembled of *Reuters News*[5] stories and divided into three parts: 203,621 training, 51,362 development, and 46,435 test tokens.

In the CoNLL'2002 shared task for Spanish and Dutch, entities are annotated using the same directives as in the English CoNLL'2003 corpus and hence comprise the same four types. The Spanish CoNLL'2002 corpus [39] consists of news wire articles from the *EFE*[6] news agency and is divided into 273,037 training, 54,837 development and 53,049 test tokens. The Dutch CoNLL'2002 corpus [39] consists of four editions of the Belgian newspaper "De Morgen" from the year 2000. The data was annotated as part of the Atranos project at the University of Antwerp. The corpus consists of 202,644 training, 37,687 development and 68,875 test tokens.

Our cross-language scenarios are based on tagged versions of Wikipedia. For English, we use the freely available resource provided by [3] as a starting point while for Spanish, we tagged the complete Spanish Wikipedia using a classifier based on the supersense tagger (SST) [9].

5.2   Experimental Setup

We use the original split of the CoNLL corpora into training, development, and test data, where the development data is used for model selection. In each experiment we compare traditional hidden Markov models and structural loss-augmented perceptrons with their extensions for learning from partially labeled data, respectively, as introduced in Sect. 4. The baselines are trained on the CoNLL training sets in the target language while their extensions additionally incorporate the automatically labeled data into the training processes. Perceptrons use 4 different groups of features, the word itself, its stem, part-of-speech, and shape/surface-clues. Features are encoded using a hash function with 18 bits,

---

[5] http://www.reuters.com/.
[6] http://efe.com/.

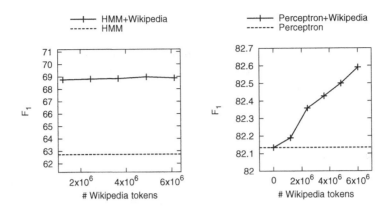

Fig. 4. Performance for English → English: HMM (left) and Perceptron (right).

allowing for a maximum of 2^{18} dimensions. We report on averages over 10 repetitions where performance is computed on the respective CoNLL test split in the target language.

5.3   Mono-Language: English → English

The goal of the mono-lingual experiment is to study an ideal scenario where every entity is trivially mapped to itself instead of applying the cross-language scenario. By doing so, every detected entity is preserved and does not have to be discarded because of missing language links. Table 7 shows some descriptive statistics of the obtained corpus. As expected, entity annotations are sparser in the automatically generated corpora compared to the CoNLL training set because of false negatives as the automatically generated corpus is only partially labeled.

Using our procedure we obtain an automatically generated corpus that is about 6,000 times larger than the CoNLL training set. To assess the importance of the size of the additional sample, we randomly sample the generated corpus into smaller subsets.

Figure 4 shows the F1 performance for varying sizes of additional Wikipedia data for hidden Markov models (left) and structural perceptrons (right). Both algorithms perform significantly better than their traditional counterparts. The HMM+Wikipedia however cannot benefit from an increasing number of additional sentences due to the limited feature representation by point-distributions. By contrast, the Wikipedia enhanced perceptron uses a much richer set of features and clearly improves its performance in terms of the number of available additional data. The improvement is marginal but significant.

Table 8. Descriptive statistics for the English → Spanish and English → Dutch corpora

	Spanish		Dutch	
	CoNLL	Wikipedia	CoNLL	Wikipedia
Tokens	264,715	257,736,886	202,644	139,404,668
Examples	8,323	10,995,693	15,806	8,399,068
Tokens per example	31.81	23.44	12.82	16.60
Entities	18,798	1,837,015	13,344	8,578,923
Entities per example	2.26	0.17	0.84	1.02
Examples with entity	74.48 %	12.15 %	46,49 %	46.22 %
MISC	11.56 %	10.59 %	25.01 %	16.16 %
PER	22.99 %	35.56 %	35.35 %	12.23 %
ORG	39.31 %	23.15 %	15.60 %	50.48 %
LOC	26.14 %	30.70 %	24.04 %	21.13 %

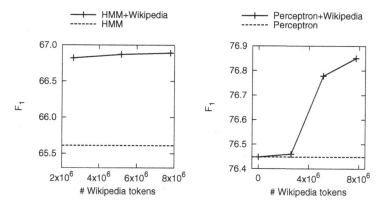

Fig. 5. Performance for English → Spanish: HMM (left) and Perceptron (right).

5.4   Cross-Language Experiments

In this section, we present results on the cross-language experiments, English → Spanish, English → Dutch and finally Spanish → English. The data generation follows the protocol described in Sect. 3.

Table 8 compares the CoNLL and the automatically generated corpora from Wikipedia for Dutch and Spanish. As before, the generated corpora are several orders of magnitude larger than their CoNLL counterparts in all respects, ranging from the number of tokens and examples to NERC annotations. Interestingly, the Spanish data has fewer entities per example and a slightly different NERC distribution than Dutch which shows a larger difference in the NERC label distribution.

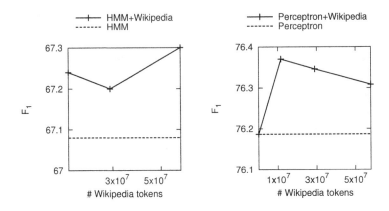

Fig. 6. Performance for English → Dutch: HMM (left) and Perceptron (right).

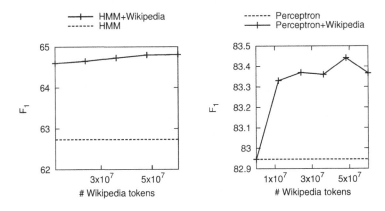

Fig. 7. Performance for Spanish → English: HMM (left) and Perceptron (right).

Figure 5 shows our empirical findings for the cross-language scenario English → Spanish. Although the differences are not as striking as in the monolingual experiment, the results reflect the same trend. Again, both Wikipedia enhanced methods consistently outperform the regular HMMs and perceptrons. While the HMM+Wikipedia hardly benefits from adding more partially labeled data, the performance of the perceptron+Wikipedia jumps for $4 \times 10e6$ additional tokens; the absolute increase is again marginal but significant.

Figure 6 details results for cross-language from English to Dutch. While the HMM shows a similar behavior as for English to Spanish, the perceptron clearly suffers from including too many unlabeled examples. The last experiment studies the cross-language scenario from Spanish to English. Since English is the biggest Wikipedia and English NLP tools are usually more accurate, using English Wikipedia as the source language seems to be a natural choice for cross-lingual NERC. Nevertheless, Fig. 7 shows the results for the uncommon Spanish → English setting.

Table 9. Descriptive statistics for the Spanish → English corpora.

	CoNLL	Wikipedia
Tokens	203,621	1,205,137,774
Examples	14,041	57,113,370
Tokens per example	14.5	21.10
Entities	23,499	11,917,106
Entities per example	1.67	0.67
Examples with entity	79.28 %	41.65 %
MISC entities	14.63 %	19.42 %
PER entities	28.08 %	12.86 %
ORG entities	26.89 %	33.87 %
LOC entities	30.38 %	33.85 %

Both methods perform as expected and exhibit the already observed slight increase in performance when more partially labeled data is added. While the HMMs are clearly outperformed by the ones trained on the mono-lingual English → English data, the perceptron surprisingly increases the performance of its single-language peer. We assume that the Wikipedia language links act like a filter for ambiguous entities so that the final bi-lingual corpus contains less noise than the mono-language data. As a consequence, the corpus generated by the cross-language approach reflects the true distribution of entities in English better than the mono-lingual counterpart where every single entity is preserved (Table 9).

6   Conclusions

We studied cross-language named entity recognition and classification (NERC) and presented an automatic approach to generate partially annotated corpora automatically from Wikipedia. Our method consisted of three stages. Firstly, we assigned the NERC tags to Wikipedia entries in the source language. Secondly, we exploited Wikipedia language links to translate entries into the desired target language. Thirdly, we generated a partially labeled corpus by annotating sentences from Wikipedia in the target language.

We devised simple extensions of hidden Markov models and loss-augmented perceptrons to learn from the partially annotated data. The data generation as well as the proposed extensions to the traditional learning algorithms were orthogonal to state-of-the-art approaches and could be easily included in any structural prediction model such as structural support vector machines and conditional random fields.

Our empirical results showed that using the automatically generated corpus as additional data is beneficial and leads to more accurate predictions than off-the-shelf methods. The observed improvements in performance were marginal but significant. We remark that NERC is mandatory for high-level text processing and that small improvements might have a large impact on higher-level applications as errors accumulate across the processing pipeline.

Future work will extend the presented figures with results for more partially labeled data and address the impact of the number of cross-language links of Wikipedia and the assignment of the labels to Wikipedia entries. We also intend to exploit the context of Wikipedia entities given by the link structure as an alternative denoising step. Although we focused on NERC as underlying task, our approach is generally applicable and can be straight-forwardly adapted to other NLP tasks including word sense disambiguation and part-of-speech tagging so that another interesting line of research is to extend our method to other sequential tasks.

References

1. Altun, Y., McAllester, D., Belkin, M.: Maximum margin semi-supervised learning for structured variables. In: Advances in Neural Information Processing Systems (2006)
2. Altun, Y., Tsochantaridis, I., Hofmann, T.: Hidden Markov support vector machines. In: Proceedings of the International Conference on Machine Learning (2003)
3. Atserias, J., Zaragoza, H., Ciaramita, M., Attardi, G.: Semantically annotated snapshot of the english wikipedia. In: Proceedings of the Sixth International Language Resources and Evaluation (LREC 2008), Marrakech, Morocco. European Language Resources Association (ELRA), May 2008
4. Augenstein, I., Maynard, D., Ciravegna, F.: Relation extraction from the web using distant supervision. In: Janowicz, K., Schlobach, S., Lambrix, P., Hyvönen, E. (eds.) EKAW 2014. LNCS, vol. 8876, pp. 26–41. Springer, Heidelberg (2014)
5. Baum, L.E., Petrie, T., Soules, G., Weiss, N.: A maximization technique occurring in the statistical analysis of probabilistic functions of markov chains. Ann. Math. Stat. **41**(1), 164–171 (1970)
6. Brefeld, U., Scheffer, T.: Semi-supervised learning for structured output variables. In: Proceedings of the International Conference on Machine Learning (2006)
7. Cao, L., Chen, C.W.: A novel product coding and recurrent alternate decoding scheme for image transmission over noisy channels. IEEE Trans. Commun. **51**(9), 1426–1431 (2003)
8. Chapelle, O., Schölkopf, B., Zien, A.: Semi-supervised Learning. MIT Press, Cambridge (2006)
9. Ciaramita, M., Altun, Y.: Broad-coverage sense disambiguation and information extraction with a supersense sequence tagger. In: Proceedings of the Conference on Empirical Methods in Natural Language Processing (EMNLP) (2006)
10. Collins, M.: Discriminative reranking for natural language processing. In: Proceedings of the International Conference on Machine Learning (2000)
11. Collins, M., Ranking algorithms for named-entity extraction: boosting and the voted perceptron. In: Proceedings of the Annual Meeting of the Association for Computational Linguistics (2002)
12. Cucerzan, S., Yarowsky, D.: Bootstrapping a multilingual part-of-speech tagger in one person-day. In: Proceedings of CoNLL 2002, pp. 132–138 (2002)

13. Dempster, A.P., Laird, N.M., Rubin, D.B.: Maximum likelihood from incomplete data via the EM algorithm. J. Roy. Stat. Soc. **39**(1), 1–38 (1977)
14. Dietterich, T.G.: Machine learning for sequential data: a review. In: Caelli, T.M., Amin, A., Duin, R.P.W., Kamel, M.S., de Ridder, D. (eds.) SPR 2002 and SSPR 2002. LNCS, vol. 2396, pp. 15–30. Springer, Heidelberg (2002)
15. Fernandes, E.R., Brefeld, U.: Learning from partially annotated sequences. In: Gunopulos, D., Hofmann, T., Malerba, D., Vazirgiannis, M. (eds.) ECML PKDD 2011, Part I. LNCS, vol. 6911, pp. 407–422. Springer, Heidelberg (2011)
16. Ferrández, S., Toral, A., Ferrández, Ó., Ferrández, A., Muñoz, R.: Exploiting wikipedia and eurowordnet to solve cross-lingual question answering. Inf. Sci. **179**(20), 3473–3488 (2009)
17. Forney, G.D.: The Viterbi algorithm. Proc. IEEE **61**(3), 268–278 (1973)
18. Gabrilovich, E., Markovitch, S.: Computing semantic relatedness using wikipedia-based explicit semantic analysis. In: Veloso, M.M. (ed.) IJCAI, pp. 1606–1611. Morgan Kaufmann Publishers Inc., San Francisco (2007)
19. Hammersley, J.M., Clifford, P.E.: Markov random fields on finite graphs and lattices. Unpublished manuscript (1971)
20. Juang, B., Rabiner, L.: Hidden Markov models for speech recognition. Technometrics **33**, 251–272 (1991)
21. Kazama, J., Torisawa, K.: Exploiting Wikipedia as external knowledge for named entity recognition. In: Proceedings of the 2007 Joint Conference on Empirical Methods in Natural Language Processing and Computational Natural Language Learning (EMNLP-CoNLL), pp. 698–707, Prague, Czech Republic, June 2007. Association for Computational Linguistics
22. Lafferty, J., Liu, Y., Zhu, X., Kernel conditional random fields: Representation, clique selection, and semi-supervised learning. Technical Report CMU-CS-04-115, School of Computer Science, Carnegie Mellon University (2004)
23. Lafferty, J., McCallum, A., Pereira, F., Conditional random fields: probabilistic models for segmenting and labeling sequence data. In: Proceedings of the International Conference on Machine Learning (2001)
24. Lafferty, J., Zhu, X., Liu, Y., Kernel conditional random fields: representation and clique selection. In: Proceedings of the International Conference on Machine Learning (2004)
25. Lee, C., Wang, S., Jiao, F., Greiner, R., Schuurmans, D.: Learning to model spatial dependency: Semi-supervised discriminative random fields. In: Advances in Neural Information Processing Systems (2007)
26. Liao, W., Veermamachaneni, S.: A simple semi-supervised algorithm for named entity recognition. In: Proceedings of the NAACL HLT Workshop on Semi-supervised Learning for Natural Language Processing (2009)
27. Màrquez, L., de Gispert, A., Carreras, X., Padró, L.: Low-cost named entity classification for catalan: exploiting multilingual resources and unlabeled data. In: Proceedings of the ACL 2003 Workshop on Multilingual and Mixed-language Named Entity Recognition, pp. 25–32, Sapporo, Japan, July 2003. Association for Computational Linguistics
28. McAllester, D., Hazan, T., Keshet, J.: Direct loss minimization for structured prediction. In: Advances in Neural Information Processing Systems (2010)
29. McCallum, A., Freitag, D., Pereira, F.: Maximum entropy Markov models for information extraction and segmentation. In: Proceedings of the International Conference on Machine Learning (2000)
30. Mihalcea, R.: Using wikipedia for automatic word sense disambiguation. In: Proceedings of NAACL HLT 2007, pp. 196–203 (2007)

31. Mika, P., Ciaramita, M., Zaragoza, H., Atserias, J.: Learning to tag and tagging to learn: a case study on wikipedia. IEEE Intell. Syst. **23**, 26–33 (2008)

32. Nadeau, D., Sekine, S.: A survey of named entity recognition and classification. Linguisticae Investigationes **30**(1), 3–26 (2007). John Benjamins Publishing Company

33. Nothman, J., Murphy, T., Curran, J.R.: Analysing wikipedia and gold-standard corpora for ner training. In: EACL 2009: Proceedings of the 12th Conference of the EuropeanChapter of the Association for Computational Linguistics, pp. 612–620, Morristown, NJ, USA (2009). Association for Computational Linguistics

34. Overell, S., Sigurbjörnsson, B., van Zwol, R.: Classifying tags using open content resources. In: WSDM 2009: Proceedings of the Second ACM International Conference on Web Search and Data Mining, pp. 64–73. ACM, New York (2009)

35. Sebastian, P., Mirella, L.: Cross-linguistic projection of role-semantic information. In: HLT/EMNLP. The Association for Computational Linguistics (2005)

36. Rabiner, L.R.: A tutorial on hidden Markov models and selected applications in speech recognition. Proc. IEEE **77**(2), 257–286 (1989)

37. Richman, A.E., Schone, P.: Mining wiki resources for multilingual named entity recognition. In: Proceedings of ACL 2008: HLT, pp. 1–9, Columbus, Ohio, June 2008. Association for Computational Linguistics

38. Ruiz-casado, M., Alfonseca, E., Castells, P.: Automatising the learning of lexical patterns: an application to the enrichment of wordnet by extracting semantic relationships from wikipedia. J. Data Knowl. Eng. **61**, 484–499 (2007)

39. Tjong Kim Sang, E.F.: Introduction to the CoNLL-2002 shared task: language-independent named entity recognition. In: COLING-2002: Proceedings of the 6th Conference on Naturallanguage Learning, pp. 1–4, Morristown, NJ, USA (2002). Association for Computational Linguistics

40. Tjong Kim Sang, E.F., De Meulder, F.: Introduction to the CoNLL-2003 shared task: Language-independent named entity recognition. In: Proceedings of CoNLL-2003, pp. 142–147 (2003)

41. Scheffer, T., Wrobel, S.: Active hidden Markov models for information extraction. In: Proceedings of the International Symposium on Intelligent Data Analysis (2001)

42. Schwarz, R., Chow, Y.L.: The n-best algorithm: An efficient and exact procedure for finding the n most likely hypotheses. In: Proceedings of the IEEE International Conference on Acoustics, Speech and Signal Processing (1990)

43. Snyder, B., Barzilay, R.: Cross-lingual propagation for morphological analysis. In: Fox, D., Gomes, C.P. (eds.) AAAI, pp. 848–854. AAAI Press, Menlo Park (2008)

44. Suzuki, J., Isozaki, H.: Semi-supervised sequential labeling and segmentation using giga-wordscale unlabeled data. In: Proceedings of ACL 2008: HLT (2008)

45. Taskar, B., Guestrin, C., Koller, D.: Max-margin Markov networks. In: Advances in Neural Information Processing Systems (2004)

46. Toral, A., Muñoz, R., Monachini, M.: Named entity wordnet. In: Proceedings of the Sixth International Language Resources and Evaluation (LREC 2008), Marrakech, Morocco. European Language Resources Association (ELRA), May 2008

47. Tsochantaridis, I., Joachims, T., Hofmann, T., Altun, Y.: Large margin methods for structured and interdependent output variables. J. Mach. Learn. Res. **6**, 1453–1484 (2005)

48. Wu, Y., Zhao, J., Xu, B., Yu, H.: Chinese named entity recognition based on multiple features. In: HLT 2005: Proceedings of the conference on Human Language Technology and Empirical Methods in Natural Language Processing, pp. 427–434, Morristown, NJ, USA (2005). Association for Computational Linguistics

49. Xu, L., Wilkinson, D., Southey, F., Schuurmans, D.: Discriminative unsupervised learning of structured predictors. In: Proceedings of the International Conference on Machine Learning (2006)
50. Yarowsky, D., Ngai, G.: Inducing multilingual pos taggers and np bracketers via robust projection across aligned corpora. In: NAACL (2001)
51. Zien, A., Brefeld, U., Scheffer, T.: Transductive support vector machines for structured variables. In: Proceedings of the International Conference on Machine Learning (2007)
52. Zinkevich, M., Weimer, M., Smola, A., Li, L.: Parallelized stochastic gradient descent. In: Advances in Neural Information Processing Systems, vol. 23 (2011)

On the Predictive Power of Web Intelligence and Social Media

The Best Way to Predict the Future Is to *tweet* It

Nathan Kallus[✉]

Massachusetts Institute of Technology,
77 Massachusetts Ave E40-149, Cambridge, MA 02139, USA
kallus@mit.edu
http://www.nathankallus.com

Abstract. With more information becoming widely accessible and new content created every day on today's web, more are turning to harvesting such data and analyzing it to extract insights. But the relevance of such data to see beyond the present is not clear. We present efforts to predict future events based on web intelligence – data harvested from the web – with specific emphasis on social media data and on timed event mentions, thereby quantifying the predictive power of such data. We focus on predicting crowd actions such as large protests and coordinated acts of cyber activism – predicting their occurrence, specific timeframe, and location. Using natural language processing, statements about events are extracted from content collected from hundred of thousands of open content web sources. Attributes extracted include event type, entities involved and their role, sentiment and tone, and – most crucially – the reported timeframe for the occurrence of the event discussed – whether it be in the past, present, or future. Tweets (Twitter posts) that mention an event to occur reportedly in the future prove to be important predictors. These signals are enhanced by cross referencing with the fragility of the situation as inferred from more traditional media, allowing us to sift out the social media trends that fizzle out before materializing as crowds on the ground.

Keywords: Web intelligence · Open-source intelligence · Web and social media mining · Twitter analysis · Forecasting · Event extraction · Temporal analytics · Sentiment analysis

1 Introduction

News from mainstream sources from all over the world can now be accessed online and about 500 million tweets are posted on Twitter each day with this rate growing steadily [16]. Blogs and online forums have become a common medium for public discourse and many government publications are accessible online. As more data than ever before is available on today's web, many are trying to harvest it and extract useful intelligence. Companies have turned to social media to

© Springer International Publishing Switzerland 2016
M. Atzmueller et al. (Eds.): MSM, MUSE, SenseML 2014, LNAI 9546, pp. 26–45, 2016.
DOI: 10.1007/978-3-319-29009-6_2

gauge consumer interest in and sentiment toward products and services as well as to the web at large to gather competitive intelligence – information about the external business environment that can guide strategic decisions. Defense and law-enforcement entities turn to the web to collect greater amounts of open-source intelligence (OSINT) – intelligence collected from publicly-available resources. And while more of this data is collected as intelligence and used to drive decisions, the precise predictive content of this data about future events is not clear. Here we investigate the potential of this publicly available information, with a focus on social media data, for predicting future events, quantifying predictive content using standard statistical measures. We focus on predicting crowd behavior and mass actions that are so significant that they garner wide mainstream attention from around the world, such as large and/or violent social unrest and politically motivated cyber campaigns.

The manifestation of crowd actions such as mass demonstrations often involves collective reinforcement of shared ideas. In today's online age, much of this public consciousness and comings together has a significant presence online where issues of concern are discussed and calls to arms are publicized. The Arab Spring is oft cited as an example of the new importance of online media in the formation of mass protests [8]. While the issue of whether mobilization occurs online is highly controversial, that nearly all crowd behavior in the internet-connected world has some presence online is not. The public information on the web that future crowds may now be reading and reacting to, or in fact posting themselves on social media, can offer glimpses into the formation of this crowd and the action it may take. Because crowd actions are events perpetrated by human actions, they are in a way endogenous to this system, enabling prediction.

But while all this information is in theory public and accessible and could lead to important insights, gathering it all and making sense of it is a formidable task. Here we use data collected by Recorded Future (www.recordedfuture.com). Scanning over 300,000 different open content web sources in 7 different languages and from all over the world, mentions of events—in the past, current, or said to occur in future—are continually extracted at a rate of approximately 50 extractions per second. Using natural language processing, the type of event, the entities involved and how, and the timeframe for the event's occurrence are resolved and made available for analysis. With such a large dataset of what is being said online ready to be processed by a computer program, the possibilities are infinite. For one, as shown herein, the gathering of crowds into a single action can often be seen through trends appearing in this data far in advance. While validating the common intuition that data on social media (beyond mainstream news sources) are able to predict major events, this quantifies the predictive power of web intelligence and social media data.

We study the cases of predicting mass and/or violent unrest by a location of interest and of politically motivated cyber campaigns by perpetrator or target. We use historical data of event mentions online to forecast the occurrence of these in the future. In particular, forward-looking discussions on social media of events yet to take place prove the most useful for prediction. We can make these predictions about specific future timeframes with high accuracy. We demonstrate this by studying the accuracy of our predictions on a large test set and by investigating in particular the case of the 2013 Egyptian coup d'état and how well we were able to foresee the unrest surrounding it.

The data used in this study has been made available at http://www.nathankallus.com/PredictingCrowdBehavior/.

1.1 Structure

We review some of the relevant literature in Sect. 1.2. We then discuss in Sect. 2 a motivating example of an event we would have liked to predict and what sort of predictive signals may exist on the web and in social media that could have helped us predict it. With this as motivation, we review how mentions of events, entities, and times are extracted from a wide breadth of sources to support prediction in Sect. 3. In Sect. 4, we use this data to develop a predictive mechanism to predict significant protests in specific future time frames both by country and by city. We consider the specific case of the 2013 Egyptian coup d'état in Sect. 5. In Sect. 6, we consider finding more general patterns in the data, motivating an application of the naïve Bayes classifier to high-dimensional sequence mining in massive datasets, which we use to forecast cyber attack campaigns by target or perpetrator. We offer some concluding remarks in Sect. 7.

1.2 Relevant Literature

Much recent work has used various online media to glean insight into future consumer behavior. In [2] the authors mine Twitter for insights into consumer demand with an application to forecasting box office ticket sales. In [10] the authors use blog chatter captured by IBM's WebFountain [9] to predict Amazon book sales. These works are similar to this one in that they employ very large data sets and observe trends in crowd behavior by huge volumes. Online web searches have been used to describe consumer behavior, most notably in [4,7], and to predict movements in the stock market in [5].

Beyond consumer behavior, Twitter data has received particular attention as to its predictive capacity in contexts ranging from finance to politics. In [3], the authors show that including the general "mood" on Twitter in stock-predictive models reduces error in predicting the Dow Jones Industrial Average by 6%. On a negative note, in [6], the author reports on the failings of Twitter data analysis to consistently predict national democratic elections. Both of these studies consider predicting the cumulative effect of many actions taken by a whole social system at once (the market buying and selling and the electorate voting). On the other hand, we seek to predict the occurence of singular events, traces of whose collaborative conception may have been left on Twitter – studying such online interactions to reveal group behavior is sometimes called social or community intelligence [19]. We also cross reference our Twitter data with data from more traditional media, which has the effect of grounding some of the volatility of Twitter.

Other work has focused only on traditional media. In [15] the authors study correlations between singular events with occurrence defined by coverage in the *New York Times*. By studying when does one target event ensue another specified event sometime in the future, the authors discover truly novel correlations between events such as between natural disasters and disease outbreaks. Here we are interested in the power of much larger, more social, and more varied datasets in pointing out early trends in endogenous processes (actions by people discussed by people) that can help predict all occurrences of an event and pinning down *when* they will happen, measuring performance with respect to each time window for prediction. One example of the importance of a varied dataset that includes both social media and news in Arabic is provided in the next section. We here seek to study the predictive power of such web intelligence data and not simply the power of different machine learning algorithms (e.g. random forests vs. SVMs). Therefore we present only the learning machine that performed best on the training data and compare it to a data-poor maximum-likelihood random-walk predictor.

2   Predictive Signals in Web Intelligence

We begin by exemplifying anecdotally the precursory signals that exist in web intelligence for large protests. On Sunday June 9, 2013 a Beirut protest against Hezbollah's interference in Syria turned violent when clashes with Hezbollah supporters left one protester dead [13]. The story was widely reported on June 9 including in Western media, attracting more mainstream news attention than any protest event in Lebanon in over a year marking it as a *significant protest*.

But not only were there signs that the protest would occur before it did, there were signs it may be large and it may turn violent. The day before, Algerian news source *Ennahar* published an article with the headline "Lebanese faction organizes two demonstrations tomorrow rejecting the participation of Hezbollah in the fighting in Syria" (translated from Arabic using Google Translate). There was little other preliminary mainstream coverage and no coverage (to our knowledge) appeared in mainstream media outside of the Middle-East-North-Africa (MENA) region or in any language other than Arabic. Moreover, without further context there would be little evidence to believe that this protest, if it occurs at all, would become large enough or violent enough to garner mainstream attention from around the world.

However, already by June 5, four days earlier, there were many Twitter messages calling people to protest on Sunday, saying "Say no to #WarCrimes and demonstrate against #Hezbollah fighting in #Qusayr on June 9 at 12 PM in Downtown #Beirut" and "Protest against Hezbollah being in #Qusair next Sunday in Beirut." In addition, discussion around protests in Lebanon has included particularly violent words in days prior. A June 6 article in *TheBlaze.com* reported, "Fatwa Calls For Suicide Attacks Against Hezbollah," and a June 4 article in the pan-Arabian news portal *Al Bawaba* reported that, "Since the revolt in Syria, the security situation in Lebanon has deteriorated." A May 23 article in the *Huffington Post* mentioned that, "The revolt in Syria has exacerbated tensions in Lebanon, which . . . remains deeply divided." Within this wider context, understood through the lens of web-accessible public information such as mainstream reporting from around the world and social media, there was a significant likelihood that the protest would be large and turn violent. These patterns persist across time; see Fig. 1.

It is these predictive signals that we would like to mine from the web and employ in a learning machine in order to forecast significant protests and other crowd behavior. In this case, it is critical that we spread a wide net on many sources so to catch mentions in non-Western media and foreign-language Tweets along with mentions in media (such as Reuters) and languages (such as English) with a more global reach.

3 Event Extraction

To quantify these signals we will look at time-stamped event-entity data. The data harvesting process extracts mentions of events from the plethora of documents continually gathered from the over 300,000 sources being monitored. An important aspect is that the event mentions are tagged with the time range in which the event is said to occur in the mention so that forward-looking statements, such as plans to protest, can be directly tied to a future time and place. Event extractions are done in Arabic, English, Farsi, French, Russian, Spanish, and Simplified and Traditional Chinese.

There are several elements in the event, entity, and time extraction process. For each document mined from the web, an ensemble of off-the-shelf natural language processing tools are used to extract tokens (lemma, root, stem, and part of speech) and entities. Entities extracted by each tool are then combined

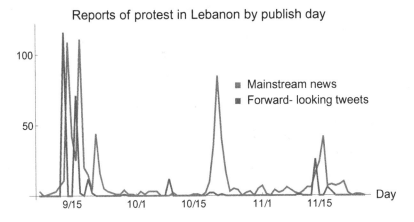

Fig. 1. Number of same-day news reports of protest in Lebanon (red) and tweets mentioning protests said to occur over the next three days (blue). Red spikes often follow blue spikes or otherwise have a recognizable convex ramp-up (Color figure online).

and resolved into a database of canonical entities for disambiguation. Documents are categorized by topic. Entities that are not agreed upon by the various tools and are far-fetched given the topic are rejected. Ontologies of structured entity relationships constructed from online sources (such as DBpedia) are used to guide filtering and provide a gazetteer for additional entity extraction.

Given the set of filtered entities, a statement extractor links these entities to events stated in the document. Events are again extracted using an ensemble method of off-the-shelf tools and a custom made tool that relies on the above tokenization. Each event from the various extraction sources is matched up with a particular text fragment from the document that best represents it. Next, n-grams in the document are matched against phrase lists organized by sentiment or tone and the fraction of these in the fragment is recorded.

All time statements made in the document are separately extracted. The tokenization of the document is parsed by a dependency grammar using the data-driven parser MaltParser [12] to construct a dependency graph of the document. This is used to find time statements, both relative (e.g. "next summer") and absolute. Comprehending these in machine time is based on several contextual cues. Cultural and regional categorizations are extracted from the document to inform such things as relevant hemisphere for seasons, which is the first day of the week, standard date formats (month first or day first), and timezone. Moreover, contexts such as publishing date are noted. Using these, all time statements made in the document are converted to standardized time-stamps, with specificity varying from second to year. Event mentions are then matched up with the most relevant time statement to the event statement based on sentence dependency.

In all, events are marked by type of event, time range of event, entities involved, role of entities involved, entities mentioned, sentiment and tone, and origin. Some post-processing is done on the event-entity level to further improve quality based on special curated ontologies. For example, known hacker groups such as Anonymous constitute one such ontology and if mentioned in a cyber event but not clearly as the perpetrator these are assumed to be so nonetheless. Similarly, impossible events are rejected. For example, one ontology keeps track of death dates of people based primarily on information harvested from Wikipedia and assists in rejecting the event "Marco Polo will travel to China in 2015" as impossible because Marco Polo is dead.

The precision of event extraction is measured by workers on the Amazon Mechanical Turk (mturkers). To test the precision of time-stamping, for each language language-specific Human Intelligence Tasks (HITs) are constructed for a random sample. Each HIT consists of the text fragment, the extracted time-stamp, and the question "is extracted time right or wrong?" and is given to three mturkers. It is declared successful if at least two answer "right." For example, the precision of time-stamping in both English and Spanish is measured at 93 %, in Arabic at 90 %, and in Simplified Chinese at 82 %. The precision of the event extraction is measured similarly by type. Protest events in English come in at 84 %. Malware threat events in English come in at 96 % and in Simplified Chinese at 90 %.

Table 1. Protest event mentions in the corpus.

Country	All	Twitter
Afghanistan	60918	27655
Bahrain	246136	177310
Egypt	944998	397105
Greece	122416	70521
India	491475	274027
Indonesia	34007	17120
Iran	118704	53962
Italy	65569	43803
Jordan	35396	19369
Lebanon	44153	23394
Libya	162721	69437
Nigeria	70635	38700
Pakistan	289643	213636
Saudi Arabia	39556	13670
Sudan	28680	13654
Syria	212815	79577
Tunisia	99000	27233
Yemen	70583	16712

4 Predicting Significant Protests

We now turn our attention to the use of this event data to the prediction of significant protests around the world. Our first forecasting question will revolve around predicting significant protests on the country level and considering that country alone. That is, a significant protest is one that receives much more same-day mainstream reporting than is usual for that country. So while most days a business with supply chain operations in Egypt operate under the usual volatile circumstances (since 2011) of Egypt—certainly more volatile than, say, Jordan and receiving much more attention for it—they are interested in receiving advance notice of protests that are going to be larger and more dangerous than the ordinary for Egypt. The same for another milieu. At the same time, we will use past patterns from other countries to inform the prediction mechanism when making a prediction about one country. In fact, the prediction mechanism will not be knowledgeable of the particular country in question but instead just the *type* of country, quantified by cluster membership.

We restrict to a selection of 18 countries: Afghanistan, Bahrain, Egypt, Greece, India, Indonesia, Iran, Italy, Jordan, Lebanon, Libya, Nigeria, Pakistan, Saudi Arabia, Sudan, Syria, Tunisia, and Yemen. We will consider all protest event mentions in any of these countries being published any time between January 1, 2011 and July 10, 2013 as the event mention corpus. January 1, 2011 up to March 5, 2013 will serve for supervised training (and validation by cross-validation) and since March 6, 2013 up to July 10, 2013 will serve for test and performance scoring. Let

$$M_{cs}(i,j) = \begin{array}{l} \text{Number of event mentions of protest} \\ \text{in country } c \text{ taking place on day } j \\ \text{extracted from documents published on} \\ \text{day } i \text{ from sources of type } s \end{array}$$

We tabulate some totals of these numbers over the whole event mention corpus in Table 1. For example, nearly one million mentions of a protest event in Egypt occur in the data, over a third of a million on Twitter.

4.1 The Ground Set

For each country, the protests we are interested in forecasting are those that are significant enough to garner more-than-usual real-time coverage in mainstream reporting for the country. That is, there is a significant protest in country c on day i if $M_{c,\text{Mainstream}}(i,i)$ is higher than usual for country c. Since new sources are being added daily by the hundreds or thousands to the Recorded Future source bank, there is a heterogeneous upward trend in the event mention data and what is more than usual in counts changes. To remove this trend we normalize the mention counts by the average volume in the trailing three months. That is, we let

$$M'_{cs}(i, i+k) = \frac{M_{cs}(i, i+k)}{\frac{1}{|\text{Countries}| \times 90} \sum_{c' \in \text{Countries}} \sum_{j=i-90}^{i-1} M_{c's}(j, j+k)}$$

where Countries is the 18 countries. Next we define the training-set average of same-day mainstream reporting

$$\overline{M'_c} = \frac{1}{|\text{Train}|} \sum_{i \in \text{Train}} M'_{c,\text{Mainstream}}(i, i)$$

where Train denotes the set of days in the training set.

Moreover, to smooth the data we consider a three-day moving average. Then we say, by definition, that a significant protest in country c (and relative to country c) occurs during the days $i - 1, i, i + 1$ if

$$M''_c(i) = \frac{1}{3} \sum_{j=i-1}^{i+1} \frac{M'_{c,\text{Mainstream}}(j, j)}{\overline{M'_c}} \geq \theta$$

is larger than a threshold θ. The threshold is chosen so to select only significant protests. By inspecting the data's correspondence to the largest protests, we set $\theta = 2.875$ (which is also nearly the 94th percentile of the standard exponential

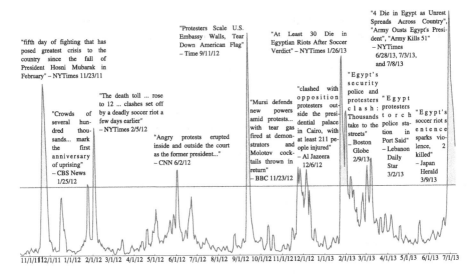

Fig. 2. Normalized count of mainstream reports $M_c''(i)$ in Egypt with annotations for stretches above $\theta = 2.875$ (red line) (Color figure online).

distribution). Overall across all countries considered, this resulted in 6 % of 3-day stretches to be labeled positive, distributed mostly evenly among the countries.

An example plot of $M_c''(i)$ for Egypt is shown in Fig. 2 with annotations of top mainstream news describing the protests in each stretch of above-threshold days. Notable protest events that may provide the reader with anchor points are the 9/11 anniversary protests in 2012 (concurrent with the Benghazi attacks), the late-June protests leading up to the Egyptian coup d'état, the riots set off by soccer-fan violence in early 2012, and the riots after 30 of the fans involved were sentenced to death which also coincided with riots connected to the anniversary of the revolution in early 2013.

4.2 Scoring Protest Predictions

We are interested in predicting on each day i whether a significant protest will occur over the next three days $i + 1, i + 2, i + 3$ based on information published on or before i. That is, on each day i we wish to predict whether $M_c''(i + 2) \geq \theta$ (which depends on days $i + 1, i + 2, i + 3$).

We quantify the success of a predictive mechanism based on its balanced accuracy. Let $T_{ci} \in \{0, 1\}$, $P_{ci} \in \{0, 1\}$ respectively denote whether a significant protest occurs in country c during the days $i + 1, i + 2, i + 3$ and whether we predict there to be one. The true positive rate (TPR), also known as recall, is the fraction of positive instances ($T_{ci} = 1$) correctly predicted to be positive ($P_{ci} = 1$) and the true negative rate (TNR) is the fraction of negative instances predicted negative ($P_{ci} = 1$). The balanced accuracy (BAC) is the unweighted average of these: BAC = (TPR + TNR)/2. BAC, unlike the marginal accuracy, cannot be artificially inflated. Always predicting "no protest" without using any data will yield a high 94 % marginal accuracy but only 50 % balanced accuracy. In fact, a prediction without any relevant data will always yield a BAC of 50 % on average by statistical independence.

4.3   The Features

We now attempt to quantify the predictive signals we encountered anecdotally in Sect. 2. These features will serve as the data based on which we make predictions.

In Sect. 2 we exemplified how the violence in language surrounding discussion of protest in a country can help set the context for the danger of a future protest to get out of hand. Each event mention is rated for violent language by the fraction of n-grams in the corresponding fragment (ignoring common words) that match a phrase list. Let

$$V_{cs}(i) = \begin{array}{l}\text{Total violence rating of fragments associated with} \\ \text{event mentions of protest in country } c \text{ extracted from} \\ \text{documents published on day } i \text{ from sources of type } s\end{array}$$

Similarly to the normalization of event mentions due to the ever-growing source bank, we normalize this quantity as

$$V'_{cs}(i) = \frac{V_{cs}(i)}{\frac{1}{|\text{Countries}|\times 90}\sum_{c'\in\text{Countries}}\sum_{j=i-90}^{i-1}V_{c's}(j)}$$

In addition, forward-looking mentions in mainstream news and Twitter can help indicate whether a protest is planned and estimate how many might plan to attend. We have already defined $M_{cs}(i, i + k)$ which counts this data for $k \geq 1$.

In order to facilitate trans-country training, we normalize these features with respect to the series we would like to predict, $M'_{c,\text{Mainstream}}(i, i)$. Similar to the normalization of $M''_c(i)$, we normalize these features by a per-country constant coefficient $\left(\overline{M'_c}\right)^{-1}$.

For the purposes of trans-country training, we hierarchically cluster the countries using Ward's method [18] applied with the distance between two countries c, c' equal to the Kolmogorov-Smirnov uniform distance between the distribution functions of the set of training values of M'' ignoring the time dimension. That is, $d(c, c') =$

$$\sup_{z\geq 0}\left|\frac{1}{|\text{Train}|}\sum_{i\in\text{Train}}\left(\mathbb{I}\left\{z \geq M''_c(i)\right\} - \mathbb{I}\left\{z \geq M''_{c'}(i)\right\}\right)\right|$$

This distance is also a non-parametric test statistic to test the hypothesis that two samples were drawn from the same distribution. We construct $\lfloor 2\sqrt{|\text{Countries}|}\rfloor$ clusters using R function hclust [14]. We include as a feature the indicator unit vector of cluster membership of the country c associated with the instance (c, i). Thus the classifier does not know the particular country about which it is making a prediction, just its type as characterized by this clustering.

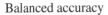

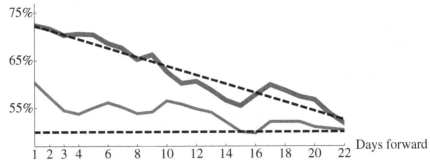

Fig. 3. BAC for predicting protests by distance into the future using the random forest and the full data-set (blue and trend in dashed black) and the data-poor predict-like-today heuristic (red) (Color figure online).

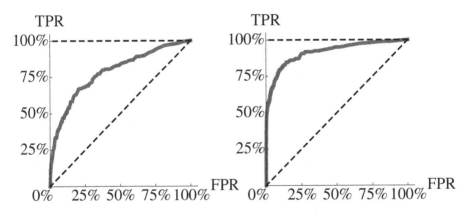

Fig. 4. Achievable TPR-FPR rates as we vary the classification threshold for (left) country-level predictions with a relative significance scale and (right) city-level predictions with an absolute significance scale.

For each instance (c, i) we also include as features the ten most recent days of same-day reporting on protest in c,

$$\frac{M'_{c,\text{Mainstream}}(i, i)}{\overline{M'_c}}, \ldots, \frac{M'_{c,\text{Mainstream}}(i - 9, i - 9)}{\overline{M'_c}}$$

along with the two most recent differences of these values. We also include the violence rating in recent mainstream reporting as the cumulative partial sums of the values

$$\frac{V'_{c,\text{Mainstream}}(i)}{\overline{M'_c}}, \ldots, \frac{V'_{c,\text{Mainstream}}(i - 9)}{\overline{M'_c}}$$

Next we include the counts of mentions of protests said occur over the next three days, published either in mainstream news or in Twitter over the ten recent days. We incorporate this feature as the cumulative partial sums of the values

$$\frac{\sum_{k=1}^{3} M'_{cs}(i, i+k)}{\overline{M'_c}}, \ldots, \frac{\sum_{k=1}^{3} M'_{cs}(i-9, i+k)}{\overline{M'_c}}$$

for s = Mainstream and s = Twitter.

When predicting farther into the future, about the three days starting with the kth day from today, we push all indexes back by $k-1$ days, thus excluding any future data and maintaining the same overall length of the feature vector.

4.4 The Classifier

For the prediction mechanism we employ a random forest classifier trained on all data up to March 5, 2013. We use the R library randomForest [11]. The parameters are left at their default values (as in version 4.6–10 of the library) as set in the library. For example, the defaults dictate that the forest have 500 trees each trained on $\lfloor \sqrt{\#\text{features}} \rfloor$ randomly chosen features.

We tune only the threshold required for a positive prediction. If the fraction of trees in the forest voting positive is at least this threshold then a positive prediction is made, otherwise negative. We tune this by four-fold cross validation over the training set using the BAC metric.

4.5 Results

We tested the trained random forest on test data March 6, 2013 to July 10, 2013. The results were TPR = **75.51 %** and TNR = **69.31 %**. BAC, the average of these two, is **72.41 %**, which constitutes a **44.8 %** reduction in balanced error from having no data. In comparison, predicting for the future the situation today, simulating the data-poor prediction possible when one nonetheless has information about today's situation (whether from being at the location or from news), has TPR = **27.04 %**, TNR = **93.74 %**, and BAC = **60.39 %**.

As we attempt to predict farther into the future our predictions become noisier and closer to the no data case as the earlier data has less bearing on the far future and there are fewer reports mentioning events to occur on the days in question. In Fig. 3 we plot the accuracy of making predictions farther into the future. For each $k \geq 1$ we re-train the random forest with the pushed-back feature vectors.

As we vary the voting threshold, we can (monotonically) trade off true positives with true negatives. We plot the range of achievable such rates for our classifier in Fig. 4 (left), using the false positive rate (FPR = $1 -$ TNR) as is common by convention. The area under the curve (AUC), here **78.0 %**, is the probability that a random positive instance has more positive votes than a random negative instance. The fraction of trees voting positive on an instance could well serve as a risk rating. By randomizing between two thresholds, we can achieve any weighted average of the rates. The rates achievable via randomization is the convex hull of the curve in Fig. 4 and has **79.2 %** under it.

4.6   Predicting Protests on the City Level

To show the breadth of this approach, we also consider an alternative forecasting question where we instead wish to predict protests on the city level and we define significance on a global scale. That is, a protest is deemed significant if it garners unusually high same-day mainstream reporting relative to all cities considered. Using a global scale, the fraction of trees voting positive serves as an absolute risk rating that is comparable between cities. The predictability of positive events also increases because knowing the city alone is no longer statistically independent of there being a positive event (for example, 23 % of positive training instances are in Cairo). The further localized data also improves predictability.

From within the countries considered previously, we choose the top cities by number of mentions of protest events. These 37 cities are Jalalabad, Kabul, and Kandahar in Afghanistan; Manama in Bahrain; Alexandria, Cairo, Port Said, and Tanta in Egypt; Athens in Greece; Hyderabad, Mumbai, and New Delhi in India; Jakarta in Indonesia; Tehran in Iran; Milan and Rome in Italy; Amman in Jordan; Beirut and Sidon in Lebanon; Benghazi and Tripoli in Libya; Abuja and Lagos in Nigeria; Islamabad, Karachi, Lahore, Peshawar, and Quetta in Pakistan; Qatif and Riyadh in Saudi Arabia; Khartoum in Sudan; Aleppo, Damascus, Deraa, Hama, and Homs in Syria; Tunis in Tunisia; and Sana'a and Taiz in Yemen. We consider the same time range and the same train-test split.

We define M' and V' as before but for these cities. Since we are interested in an absolute level of significance we no longer normalize with respect to the entity, only with respect to the cross-entity average trailing volume. Thus, while overall still only 6 % of instances are labeled positives, Cairo takes up 23 % of positive training instances and 40 % of positive test instances, while Khartoum takes up 1 % of positive train instances and has no positive test instances.

Nonetheless, as before, cities are clustered according to their set of M' training values and the classifier is not knowledgeable of the particular city in question, just its cluster membership. We use as features the unnormalized violence rating of past ten days of mainstream reporting about the city, same-day mainstream reporting level of past ten days, and the forward-looking mainstream reporting and Twitter discussion of past ten days. In addition, we include the unnormalized features of the containing country.

A random forest classifier is trained with the voting fraction threshold tuned by four-fold cross-validation to maximize balanced accuracy. Any other parameters were set to their defaults as before. Testing on March 6, 2013 to July 10, 2013, we get a true positive rate of **84.7 %** and a true negative rate of **85.7 %** yielding a balanced accuracy of **85.2 %**. The achievable rates as the voting fraction threshold is varied are shown in Fig. 4 (right). The area under the ROC curve (AUC) is **91.3 %** and the area under its convex hull is **91.9 %**.

5   The Case of the 2013 Egyptian Coup d'État

To exemplify the prediction mechanism we study predictions made for Egypt on different days about the relative future around the time of the 2013 coup

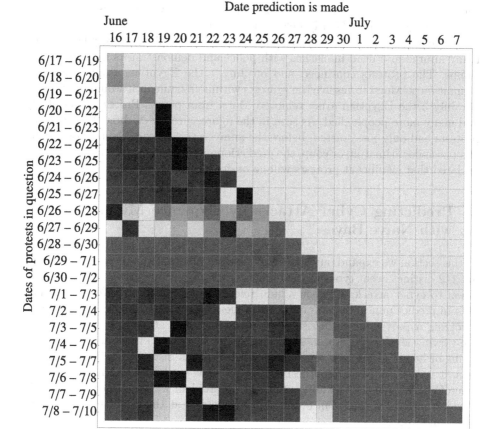

Fig. 5. Predictions of protests in Egypt around the coup d'état. Yellow to red mark positive predictions and blue to purple negative. Redder colors indicate more positive votes (Color figure online).

d'état in Egypt (Fig. 5). As shown, the days around June 30 were predicted positive with very high certainty for a long time prior, the date ranges in 6/28–7/2 being consistently predicted positive since June 6, three weeks beforehand (since June 16 onward shown in figure). Indeed, with a lot of discontent with and talk of demonstration against President Morsi's rule, many protests were anticipated for the weekend of June 30, the anniversary of Morsi's rise to the presidency. Even U.S. Secretary of State John Kerry made a statement in anticipation of protests asking for peaceful and responsible demonstration on, as he says, Saturday (June 29) and Sunday (June 30) [17]. Therefore, those days were long predicted positive with high certainty. However, already on June 28 spontaneous "warm-up" protests burst in the streets [1]. The first range to include June 28 was predicted with less certainty, especially from farther back, but was correctly predicted starting June 10 except on June 16 when it was mistakenly reported negative, just slightly below the threshold.

As we now know the protests around the anniversary did grow very large with many injured and dead in clashes with police and demonstrators from opposite camps. The protests continued and on July 1 the Egyptian army issued an ultimatum to Morsi to resolve the protests within 48 hours or it would intervene. On July 3 the Egyptian army removed Morsi from power. Protests intensified and many more people died. As seen in the figure, already on June 28 when the protests had only just started, before the anniversary and before any discussion of a possibile ultimatum or coup, the prediction mechanism had already correctly declared that significant protests will go on for the weeks to come.

6   Predicting Cyber Attacks by Sequence Mining with Naïve Bayes

In this section we expand our scope and consider all event types recorded. There are 112 distinct event types in the data ranging from protest to company acquisition to cyber attack to music album release to voting result. We wish to be able to predict unusually high numbers of mentions of a particular type of event involving a particular entity. Here we focus on cyber attacks. However, there are varying levels of "clumpiness" for the many classes of events and entities in terms of how and for how long a real-world event is discussed online. In addition, it is often hard to hypothesize *a priori* what predictive signals may exist. Therefore, in order to tackle this forecasting problem we would need to spread a wider net and consider all event interactions and at the same time allow for more smoothing.

We will therefore consider events on the week level for a given entity n^* (which could be a country, a person, an organization, a product, among many other things) and use events mentioning that entity to forecast the level of mentions of an event type of interest involving that entity next week. Let

$$M_{n^*es}(i,j) = \begin{array}{l} \text{Number of event mentions of type } e \\ \text{involving entity } n^* \text{ taking place on week} \\ j \text{ extracted from documents published} \\ \text{on week } i \text{ from sources of type } s \end{array}$$

Here we will consider source types Any, Mainstream, Social Media, and Blog. As before, we normalize this number with respect to the total event mention volume in the past 12 weeks (approximately three months, as before) in order to de-trend it as follows

$$M'_{n^*es}(i,i+k) = \frac{M_{n^*es}(i,i+k)}{\sum_{e' \in \text{EventTypes}} \sum_{j=i-12}^{i-1} M_{n^*e's}(i,i+k)}$$

This is the data we will feed to our prediction algorithm. We will consider both the mentioning over the past weeks of events taking place in that same week as well as any forward-looking mentions on a past week of events to take place next week, the week in question. As before, we will also consider the case where we must predict farther into the future, about the week after next or the one after that etc.

We will consider data starting from the first week of 2011 and up to the last week of July 2013. We will test our mechanism on April 2012 onward, training on the trailing two years (as available). Any cross-validation is done on 2011 up to March 2012.

6.1 The Ground Set

Along with an entity of interest n^*, let us fix an event type of interest e^* and a source type of interest s^*. Because we believe our data is particularly useful in predicting crowd behavior we will choose e^* accordingly. Here we will be interested in predicting politically motivated cyber campaigns so we fix $e^* = $ CyberAttack. We label as positive weeks that included cyber attack campaigns that were so impactful to generate wide attention all over with same-week mentions of cyber attack events. Therefore we fix $s^* = $ Any. We will consider n^* that are both country target entities (such as Israel) and hacktivist attacker entities (such as Anonymous).

We also fix a threshold θ and we will wish to predict on week i whether

$$M'_{n^*e^*s^*}(i+, i+) \geq \theta$$

We fix θ so that 15 % of weeks are positive.

As before, we will use balanced accuracy to score our predictive mechanism and to tune parameters by cross-validation.

6.2 High-Dimensional Sequence Mining with Naïve Bayes

Let $T_i = 1$ denote the positivity of the prediction instance on week i

$$T_i = 1 \; : \; M'_{n^*e^*s^*}(i+1, i+1) \geq \theta$$

We seek to estimate the conditional probability density conditioned on the past ℓ weeks

$$\mathbb{P}\left(T_i = t \; \middle| \; \begin{array}{c} M'_{n^*es}(i-k, i-k),\, M'_{n^*es}(i-k, i+1) \\ \text{for } e \in \text{EventTypes}, \\ s \in \text{SourceTypes}, \\ k = 0, \ldots, \ell-1 \end{array}\right) \qquad (1)$$

for $t = 0$ or 1. We use $\ell = 5$ here.

That estimate this, we seek to find the patterns of ℓ event sequences that end with our target event. Sequence mining is the discovery of commonly repeating sequences in a string over an alphabet Σ. In bioinformatics, sequence mining is applied to DNA sequences ($\Sigma = \{A, C, G, T\}$) and to amino acid sequences constituting a protein ($|\Sigma| = 20$) to find common sequences of some length. For longer strings the frequency of appearing in nature is highly concentrated.

We first bin the values of $M'_{n^*es}(i, i+k)$ into quartiles of their marginal distribution over the training data. The resulting alphabet Σ has a cardinality of $4^{2 \times \ell \times |\mathrm{SourceTypes}| \times |\mathrm{EventTypes}|}$, much larger than the training data set so that the probability function is underspecified (high-dimensional setting). At the same time, the amount of information in the training data is also quite massive. So we require a method that can smooth the density to avoid overfitting and, at the same time, tractable over a large data set.

One solution to this problem is to apply what is known as the naïve assumption to our likelihood function (1). Let $F, F' \in \{1, 2, 3, 4\}$ (indicating the binned quartile) be any two different features of the past in the conditioning in (1). Then we make the following assumption:

Conditioned on T_i, F and F' are statistically independent.

This is of course very different from assuming marginal independence. For example under the assumption, discussion of protests or of military actions in Israel in mainstream news could very often coincide with a discussion on Twitter of a planned future cyber attack against Israel because a possible future cyber attack is often a response to the former; but given that a cyber attack against Israel does occur next week, we assume the former two must occur independently from one another.

By Bayes' theorem we may decompose the probability function to a product of $\mathbb{P}(T_i = t)$ and the conditional probabilities of $M'_{n^*es}(i-k, i-k)$ and $M'_{n^*es}(i-k, i+1)$ given $T_i = t$. Estimating instead the marginal distribution of T and the conditional distributions of M' by maximum likelihood (counting co-occurrences in training data) results in the well known naïve Bayes probability estimator. This reduces the variance of the density estimator but introduces bias whenever the naïve assumption does not hold exactly.

To further relieve issues of the sparsity of positives in our data, instead of estimating the conditional probabilities by maximum likelihood, we take a Bayesian approach and assume a Dirichlet prior with symmetric concentration

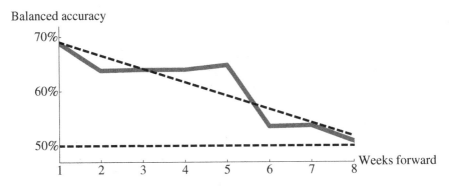

Fig. 6. In dark blue, the balanced accuracy of predicting cyber attacks against Israel by distance into the future (Color figure online).

Table 2. Accuracy of predicting cyber attacks against and by a selection of entities.

Targets	BAC	Perpetrators	BAC
Israel	68.9 %	Anonymous	70.3 %
Germany	65.4 %	AnonGhost	70.8 %
South Korea	63.1 %	LulzSec	60.6 %
United Kingdom	65.5 %	Guccifer	66.7 %

parameter α. The prior is our probabilistic belief about the value of these conditional probabilities in the absence of data. Being the conjugate prior to the categorical distribution, a Dirichlet law will also be the distribution for the posterior distribution. Estimating the parameters by maximum a posteriori likelihood is then equivalent to estimating the probabilities by counting frequencies from the data and padding these counts each by α. We estimate the marginal $\mathbb{P}\,(T_i = 1)$ using maximum likelihood because that estimation is not plagued by sparsity.

Estimating the conditional probabilities thus is then done simply by counting data which can be done exceedingly fast. For each event, source, and look-back k, one simply counts occurrences in bins of the feature and keeps such tallies separately for when $k + 1$ weeks forward (only if within the training data) has been positive or negative. The computation involved in this procedure scales as the product of the length of the data and the number of features.

To make a prediction we check whether $(1) \geq p^*$ for a chosen threshold p^*. Due to the bias introduced by making the naïve assumption, we select p^* by four-fold cross-validation on the training data (up to March 2012) to maximize balanced accuracy.

6.3   Results

We first apply this method to predicting cyber campaigns against Israel. For reference, most of these are perpetrated by groups AnonGhost and Anonymous under the banner of #OpIsrael. Testing on April 2012 to July 2013, we get a true positive rate of **70.0 %** and a true negative rate of **67.8 %** giving a balanced accuracy of **68.9 %**. Our accuracy diminishes as we try to predict farther into the future, as depicted in Fig. 6.

By inspecting the trained conditional probabilities we can see which were the most impactful features to sway our belief one way or the other. In this case, swaying our belief most toward predicting positively were if many blog and mainstream mentions of cyber attack appeared in recent weeks and if many social media mentions of protest in Israel appeared in recent weeks as well as forward-looking mentions on social media of a protest in the week to come.

We apply the same method to predicting attacks against three other country entities and to predicting campaigns perpetrated by a selection of four hacktivist groups. The results are reported in Table 2.

Where hacktivism campaigns are often reactions to developments that do not necessarily at first involve the hacktivist organization, incorporating in some way event mentions involving other entities could boost performance. However, it is not immediately clear how to do so without introducing too many redundant and obfuscating features that will result in overfitting and poor out-of-sample accuracy. Using abstractions as in [15] is one possible way to improve this method.

7   Conclusions

We presented new findings about the power of web intelligence and social media data to predict future events. The evidence presented validates and quantifies the common intuition that data on social media (beyond mainstream news sources) are able to predict major events. The scope and breadth of the data offered glimpses into tweets in foreign languages and news in far places. The confluence of all this information showed trends to come far in the future. That we found this to be true consistently across applications suggests that there is significant predictive content in web intelligence and social media data relevant to the future and the analysis of such data can provide important insights into it.

References

1. Ahram.org. Egypt warms up for a decisive day of anti- and pro-Morsi protests. www.english.ahram.org.eg/NewsContent/1/64/75483/Egypt/Politics-/Egypt-war ms-up-for-a-decisive-day-of-anti-and-proM.aspx. Accessed 25 August 2013
2. Asur, S., Huberman, B.A.: Predicting the future with social media. In: WI-IAT (2010)
3. Bollen, J., Mao, H., Zeng, X.: Twitter mood predicts the stock market. J. Comput. Sci. **2**(1), 1–8 (2011)
4. Choi, H., Varian, H.: Predicting the present with google trends. Econ. Rec. **88**(s1), 2–9 (2012)
5. Da, Z., Engelberg, J., Gao, P.: In search of attention. J. Finance **66**(5), 1461–1499 (2011)
6. Gayo-Avello, D.: No, you cannot predict elections with twitter. IEEE Internet Comput. **16**(6), 91–94 (2012)
7. Goel, S., Hofman, J.M., Lahaie, S., Pennock, D.M., Watts, D.J.: Predicting consumer behavior with web search. PNAS **107**(41), 17486–17490 (2010)
8. González-Bailón, S., Borge-Holthoefer, J., Rivero, A., Moreno, Y.: The dynamics of protest recruitment through an online network. Sci. Rep. **1**, 197 (2011)
9. Gruhl, D., Chavet, L., Gibson, D., Meyer, J., Pattanayak, P., Tomkins, A., Zien, J.: How to build a webfountain: an architecture for very large-scale text analytics. IBM Syst. J. **43**(1), 64–77 (2004)
10. Gruhl, D., Guha, R., Kumar, R., Novak, J., Tomkins, A.: The predictive power of online chatter. In: SIGKDD (2005)
11. Liaw, W.: Classification and regression by randomForest. R News **2**(3), 18–22 (2002)
12. Nivre, J., Hall, J., Nilsson, J., Chanev, A., Eryigit, G., Kübler, S., Marinov, S., Marsi, E.: MaltParser: a language-independent system for data-driven dependency parsing. Nat. Lang. Eng. **13**(2), 95–135 (2007)

13. NYTimes.com. Protester Dies in Clash That Apparently Involved Hezbollah Supporters. www.nytimes.com/2013/06/10/world/middleeast/protester-dies-in-lebanese-clash-said-to-involve-hezbollah-supporters.html. Accessed 24 August 2013
14. R Core Team. R: A Language and Environment for Statistical Computing. Vienna, Austria (2013)
15. Radinsky, K., Horvitz, E.: Mining the web to predict future events. In: WSDM (2013)
16. Telegraph.co.uk. Twitter in numbers. www.telegraph.co.uk/technology/twitter/9945505/Twitter-in-numbers.html. Accessed 25 August 2013
17. TheGuardian.com. John Kerry urges peace in Egypt amid anti-government protests. www.theguardian.com/world/video/2013/jun/26/kerry-urges-peace-egypt-protests-video. Accessed 25 August 2013
18. Ward, J.: Hierarchical grouping to optimize an objective function. J. Am. Stat. Assoc. **58**(301), 236–244 (1963)
19. Zhang, D., Guo, B., Yu, Z.: The emergence of social and community intelligence. Computer **7**, 21–28 (2011)

A Latent Space Analysis of Editor Lifecycles in Wikipedia

Xiangju Qin, Derek Greene, and Pádraig Cunningham[✉]

School of Computer Science,
University College Dublin, Belfield, Ireland
xiangju.qin@ucdconnect.ie, {derek.greene,padraig.cunningham}@ucd.ie

Abstract. Collaborations such as Wikipedia are a key part of the value of the modern Internet. At the same time there is concern that these collaborations are threatened by high levels of member withdrawal. In this paper we borrow ideas from topic analysis to study editor activity on Wikipedia over time using latent space analysis, which offers an insight into the evolving patterns of editor behaviour. This latent space representation reveals a number of different categories of editor (e.g. *Technical Experts*, *Social Networkers*) and we show that it does provide a signal that predicts an editor's departure from the community. We also show that long term editors generally have more diversified edit preference and experience relatively soft evolution in their editor profiles, while short term editors generally distribute their contribution at random among the namespaces and categories of articles and experience considerable fluctuation in the evolution of their editor profiles.

1 Introduction

With the popularity of Web 2.0 technologies, recent years have witnessed an increasing population of online peer production communities which rely on contributions from volunteers to build knowledge, software artifacts and navigational tools, such as *Wikipedia*, *Stack Overflow* and *OpenStreetMap*. The growing popularity and importance of these communities requires a better understanding and characterization of user behaviour so that the communities can be better managed, new services delivered, and challenges and opportunities detected. For instance, by understanding the general lifecycles that users go through and the key features that distinguish different user groups and different life stages, we would be able to: (i) predict whether a user is likely to abandon the community; (ii) develop intelligent task routing software in order to recommend tasks to users within the same life-stage. Moreover, the contribution and social interaction behaviour of contributors plays an essential role in shaping the health and sustainability of online communities.

Recent studies have approached the issue of modeling user lifecycles (also termed user profiles or user roles) in online communities from different perspectives. Such studies have so far focused on a separate set or combination of user properties, such as information exchange behaviour in discussion forums [6],

© Springer International Publishing Switzerland 2016
M. Atzmueller et al. (Eds.): MSM, MUSE, SenseML 2014, LNAI 9546, pp. 46–69, 2016.
DOI: 10.1007/978-3-319-29009-6_3

social and/or lexical dynamics in online platforms [7,13], and diversity of contribution behaviour in Q&A sites [8]. These studies generally employed either principle component analysis and clustering analysis to identify user profiles [6,8] or entropy measures to track social and/or linguistic changes throughout user lifecycles [7,13]. While previous studies provide insights into community composition, user profiles and their dynamics, they have limitations either in their definition of lifecycle periods (e.g. dividing each user's lifetime using a fixed time-slicing approach [7] or a fixed activity-slicing approach [13]) or in the expressiveness of user lifecycles in terms of the evolution of expertise and user activity for users and the communities over time. Specifically, they fail to capture a mixture of user interests over time.

On the other hand, in recent years, there have been significant advances in topic modeling which develops automatic text analysis models to discover latent structures from time-varying document collections. In this paper we present a latent space analysis of user lifecycles in online communities specifically Wikipedia. Our contributions are summarised as follows:

- We model the lifecycles of users based on their activity over time using dynamic topic modeling, thus complementing recent work (e.g. [7,8,13]).
- This latent space analysis reveals a number of different categories of editor (e.g. *content experts*, *social networkers*) and offers an insight into the evolving patterns of editor behaviour.
- We find that long term and short term users have very different profiles as modeled by their activity in this latent representation.
- We show that the patterns of change in user activity can be used to make predictions about the user's turnover in the community.

The rest of this paper is organized as follows. The next section provides a brief review of related work. In Sect. 3, we provide an overview of dynamic topic models and explanation of data collection. Next, we present results about latent space analysis of editor lifecycles in Wikipedia, followed by performance of churn prediction and concluding remarks.

2 Related Work

Over the last decade, works about investigating and modelling changes in user behaviour in online communities have attracted much interest among researchers. Such works have been conducted in varied contexts, including discussion forums [6], Wikipedia [12,17], beer rating sites [7], Q&A sites [8,13] and other wikis. Chan *et al.* [6] presented an automated forum profiling technique to capture and analyze user interaction behaviour in discussion forums, and found that forums are generally composed of eight behaviour types such as popular initiators and supporters. Welser *et al.* [17] examined the edit histories and egocentric network visualizations of editors in Wikipedia and identified four key social roles: substantive experts, technical editors, vandal fighters, and social networkers. Panciera *et al.* [12] studied the contribution behaviours of long-term editors and newcomers in Wikipedia and their changes over time, and found significant differences between the two groups: long-term editors start intensely, tail off a little, then maintain a relatively high level of activity over the course of their career; new users follow the same trend of the evolution but do much less work than long-term editors throughout their contributory lifespans. The studies mentioned provide insights about contributor behaviour at a macro level, but are limited in capturing the change of behaviour at a user level.

Danescu-Niculescu-Mizil *et al.* [7] examined the linguistic changes of online users in two beer-rating communities by modeling their term usages, and found that users begin with an innovative learning phase by adopting their language to the community, but then transit into a conservative phase in which they stop changing the language. Rowe [13] modeled how the social dynamics and lexical dynamics of users changed over time in online platforms relative to their past behaviour and the community-level behaviour, mined the lifecycle trajectories of users and then used these trajectories for churn prediction. Based on the diversity, motivation and expertise of contributor behaviours in five Q&A sites, Furtado *et al.* [8] examined and characterized contributor profiles using hierarchical clustering and K-means algorithms, and found that the five sites have very similar distributions of contributor profiles. They further identified common profile transitions by a longitudinal study of contributor profiles in one site, and found that although users change profiles with some frequency, the site composition is mostly stable over time. The aforementioned works provide useful insights into community composition, user profiles and their dynamics, but they are limited either in their definition of lifecycle periods (e.g. dividing each user's lifetime using a fixed time-slicing approach [7] or a fixed activity-slicing approach [13]) or in the expressiveness of user profiles in terms of the evolution of expertise and user activity for users and the communities over time [8]. Specifically, they fail to capture a mixture of user interests over time.

In recent years, there has been an increasing interest in developing automatic text analysis models for discovering latent structures from time-varying document collections. Blei and Lafferty [3] presented dynamic topic models (DTM) which make use of state space models to link the word distribution and popularity of topic over time. Wang and McCallum [15] proposed the topics over time model which employs a beta distribution to capture the evolution of topic popularity over timestamps. Ahmed and Xing [2] introduced the infinite dynamics topic models (iDTM) which can adapt the number of topics, the word distributions of topics, and the topics' popularity over time. Based on topic models, Ahmed and Xing [1] further proposed a time-varying user model (TVUM) which models the evolution of topical interests of a user while allowing for user-specific topical interests and global topics to evolve over time. Topic modeling plays a significant role in improving the ways that users search, discover and organize web content by automatically discovering latent semantic themes from a large and otherwise unstructured collection of documents. Moreover, topic modeling algorithms can be adapted to many types of data, such as image datasets, genetic data and history of user activity in computational advertising. However, to our knowledge, there exists no attempt to understand how users develop throughout their lifecycles in online communities from the perspective of topic modeling.

This study complements the previous works (e.g. [7,8,13]) by characterizing the evolution of user activity in online communities using dynamic topic modeling, learns the patterns of change in user behaviour as modeled by their activity in the latent representation and demonstrates the utility of such patterns in predicting churners.

3   Model Editor Lifecycles

In this section, we provide a brief overview about dynamic topic models that we will use to model the lifecycle of Wikipedia users and introduce how we collect the data for the evaluation.

3.1   Dynamic Topic Modeling

The primary goal of this study is to apply topic models on the evolving user activity collections in order to identify the common work archetypes[1] and to track the evolution of common work archetypes and user lifecycles in online communities. For this purpose, we employ a LDA based dynamic topic model proposed by Blei and Lafferty [3], in which the word distribution and popularity of topics are linked across time slices using state space models. First, we review the generative process of the LDA model [4], in which each document is represented as a random mixture of latent topics and each topic is characterized by a multinomial distribution over words, denoted by $Multi(\beta)$. The process to generate a document d in LDA proceeds as follows:

1. Draw topic proportions θ_d from a Dirichlet prior: $\theta_d | \alpha \sim Dir(\alpha)$.
2. For each word
   (a) Draw a topic assignment from θ_d: $z_{di} | \theta_d \sim Mult(\theta_d)$.
   (b) Draw a word w_{di}: $w_{di} | z_{di}, \beta \sim Mult(\beta_{z_{di}})$.

Where α is a vector with components $\alpha_i > 0$; θ_d represents a topic-mixing vector for document d that samples from a Dirichlet prior (i.e. $Dir(\alpha)$), each component (i.e. z_{di}) of θ_d defines how likely topic i will appear in d; $\beta_{z_{di}}$ represents a topic-specific word distribution for topic z_{di}.

LDA is not applicable to sequential models for time-varying document collections for its inherent features and defects: (1) Dirichlet distributions are used to model uncertainty about the distributions over words and (2) the document-specific topic proportions θ are drawn from a Dirichlet distribution. To remedy the first defect, Blei and Lafferty [3] chained the multinomial distribution of each topic $\beta_{t,k}$ in a state space model that evolves with Gaussian distributions, denoted as follows:

---

[1] Common work archetypes refer to the types of contribution that users make in online platforms, e.g. answering questions in Q&A sites and editing main pages in Wikipedia.

$$\beta_{t,k}|\beta_{t-1,k} \sim N(\beta_{t-1,k}, \sigma^2 I) \tag{1}$$

To amend the second defect, the same authors employed a logistic normal with mean α to capture the uncertainty over proportions and used the following dynamic model to chain the sequential structure between models over time slices [3]:

$$\alpha_t|\alpha_{t-1} \sim N(\alpha_{t-1}, \sigma^2 I) \tag{2}$$

More details on the generative process of a dynamic topic model for a sequential corpus can refer to [3]. Note that documents generated using the DTM will have a mixture of topics. In Wikipedia, this indicates that a user has diverse edit interests and edits multiple namespaces or categories of articles in a specific time period.

3.2 Data Collection

In Wikipedia, pages are subdivided into 'namespaces'[2] which represent general categories of pages based on their function. For instance, the article (or main) namespace is the most common namespace and is used to organize encyclopedia articles. In many practical applications, we might be more interested in how the actual interests of editors (in terms of the categories of Wikipedia articles they have edited) change over time. For this reason, rather than using the main namespace as one feature, we further group Wikipedia articles into clusters based on their macro-categories[3]. Because the categories for articles given by Wikipedia are generally not fine-grained, we infer the macro-categories for articles by identifying candidate categories from DBpedia[4] category graph. DBpedia is one of the best known multidomain knowledge bases which extracts structure information from Wikipedia categorisation system and forms semantic graph of concepts and relations. The association between Wikipedia categories and DBpedia concepts is defined using the **subject** property of the DCIM terms vocabulary (prefixed by **dcterms:**) [9]. A category's parent and child categories can be extracted by querying for properties **skos:broader** and **skos:broaderof**, these category-subcategory relationships create connections between DBpedia concepts. We can obtain DBpedia category graph[5] by merging all the connections among DBpedia concepts together. With the category graph available, we can identify the macro-categories for Wikipedia articles by searching for the shortest paths from the categories associated with the articles to the macro-categories in the category graph. If multiple shortest paths exist, then the article is assigned to multiple macro-categories with weights proportional to the number of paths leading to a specific macro-category. For other complex methods of labelling topics, we recommend the readers refer to [9].

---

[2] http://en.wikipedia.org/wiki/Wikipedia:Namespace.

[3] At the time we collected data for this work, there was 22 macro/top-categories: http://en.wikipedia.org/wiki/Category:Main_topic_classifications.

[4] http://dbpedia.org.

[5] The category graph is a directed one due to the nature of category-subcategory structure.

Table 1. A simple example of the dataset, where the columns "Science" and "Mathematics" refer to science and mathematics related Wikipedia articles (or science and mathematics categories of articles), and the other columns represent the corresponding namespaces used in Wikipedia.

Uname	Quarter	Science	Mathematics	Article talk	Wikipedia	Wikipedia talk	User	User talk
User A	10	650	233	2	299	33	2	81

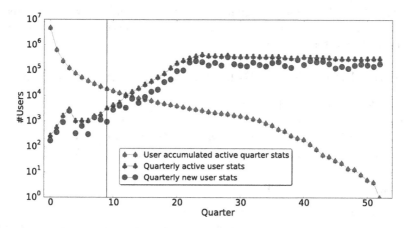

Fig. 1. Statistics about the lifespans of all registered users in Wikipedia. *Quarter* represents the number of active quarters or index of quarter; *#Users* represents the number of users for the specific stats, and is plotted on log10 scale.

Users can make edits to any namespace or article based on their edit preference. The amount of edits across all the namespaces and macro-categories can be considered as work archetypes. A namespace or macro-category can be considered as a 'term' in the vector space for document collections, the number of edits to that namespace/category is analogous to word frequency. A user's edit activity across different namespaces/categories in a time period can be regarded as a 'document'. One motivation of this study is to identify and characterize the patterns of change in user edit activity over time in Wikipedia. For this purpose, we parsed the May 2014 dump of English Wikipedia, collected the edit activity of all registered users, then aggregated the edit activity of each user on a quarterly basis (measured in 3-month period). In this way, we obtained a time-varying dataset consisting of the quarterly editing activity of all users from the inception of Wikipedia till May 2nd, 2014. The statistics about the complete dataset are as follows: 53 quarters; 28 namespaces and 22 macro-categories (or features); 6,123,823 unique registered users; 11,078,496 quarterly edit observations for all users. An example of our dataset is given in Table 1.

Figure 1 plots the statistics about the number of active quarters for all registered users. We observe that an overwhelming number of users (i.e. 4,762,685 out of 6,123,823) stayed active for only one quarter; the figures for 2, 3 and 4 quarters are 639632, 232657 and 124378, respectively. One obvious trend is that the number of users who stayed active for longer time periods is becoming smaller and smaller, indicating that Wikipedia experiences high levels of member withdrawal. Because the size of quarterly active users is relatively small from the 1st to the 9th quarter, we generated time-varying dataset using data from the 10th quarter. To avoid a bias towards behaviours most dominant in communities with larger user bases, following Furtado *et al.* [8], we randomly selected 20% of users who stayed active for only one quarter, included these users and those who were active for at least two quarters into our dataset, resulting in a time-varying dataset with 2,436,995 unique users and 6,504,726 observations. It is obvious from Fig. 1 that, the number of quarterly active users and new users increases gradually from the first quarter to 24th quarter, and becomes relatively stable after the 25th quarter.

4   Editor Lifecycles as Released by Shift in Participation

In this section, we present the analysis of editor lifecycles from two perspectives: (1) community-level evolution; and (2) user-level evolution for groups of editors. From the analysis, we identify some basic features that are useful for predicting how long a user will stay active in an online community (Sect. 5).

4.1   Community Level Change in Lifecycle

Online communities experience dynamic evolution in terms of a constantly changing user base and the addition of new functionalities to maintain vitality. For instance, online platforms like Wikipedia generally experience high levels of member withdrawal, with 60% of registered users staying only a day. Wikipedia introduced social networking elements to MediaWiki in order to attract and retain user participation[6], which is believed to increase user edit activity. As such, to understand what shape these communities, it is essential to take into account both dimensions of evolution. The dynamic topic models can accommodate both aspects of evolution, which provides valuable insights about the development of community and its users across time. To analyze the evolution of time-varying user edit activity, we ran the DTM software[7] released by Blei and Lafferty [3] with default hyperparameters and the number of topics k being 10[8].

---

[6] http://strategy.wikimedia.org/wiki/Attracting_and_retaining_participants.

[7] Available at: http://www.cs.princeton.edu/~blei/topicmodeling.html.

[8] We experimented with different number of topics $k \in [5, 45]$ with steps of 5 on the quarterly dataset using Non-negative Matrix Factorization (NMF) clustering, and then employed the mean pairwise normalized mutual information (NPMI) and mean pairwise Jaccard similarity (MPJ) as suggested by [11] to assess the coherence and generality of the topics for different ks. To cluster the quarterly data matrix efficiently, we used the fast alternating least squares variant of NMF introduced by Lin [10]. To produce deterministic results and avoid a poor local minimum, we used the Non-negative Double Singular Value Decomposition (NNDSVD) strategy [5] to choose initial factors for NMF. We found that overall, the run with 10 topics generates more coherent and general topics, and thus provides more interpretable and expressiveness results in terms of interpretation and overlapping between different topics.

Table 2. Summary of common user roles. Dominant features are in bold font.

Id	Name	Edit namespaces (Sequence indicates the importance)
1	Politics	*Politics, Geography, History, Category,* TimedText talk
2	Miscellaneous I (Technical experts)	*Wikipedia, Wikipedia talk, Portal, MediaWiki talk, MediaWiki,* Portal talk, Help, Module, Help talk
3	Miscellaneous II (Maintaince)	*Main disambution, Sports, Template,* Module talk, Professional studies
4	Society technology	*Society, Technology, Culture, Concepts, Humans, Language, Mathematics,* History, Law, Humanities
5	Miscellaneous III (Personalization)	*User, Book talk, Draft,* Book, Education Program talk, History, Humans, CategoryTalk, Concepts
6	Arts	*Arts, Humanities,* TimedText, File, Concepts, Geography
7	Communicator & coordinator	*Article talk, File,* Category talk, Template talk, File talk, Draft talk
8	People	*People, Law, Health,* Education Program
9	Social networkers	*User talk,* Wikipedia Talk, Culture, Society, Humans, Politics, File, Law
10	Nature science	*Nature, Environment, Science, Agriculture, Medicine,* Concepts, Main disambution, Society, Humans

Summary of User Roles. Table 2 presents a summary of the common user roles identified by DTM [3]. It is obvious from Table 2 that, each user role is defined by different combinations of namespaces and macro-categories. For instance, user role "Social Networkers" is featured by *user talk* namespace, then *Wikipedia Talk* namespace and *Culture* macro-category, indicating that users who are assigned to this user role spend most of their time on Wikipedia interacting with other users; user role "Miscellaneous I (Technical Experts)" is featured by *Wikipedia, Wikipedia talk, Portal, MediaWiki talk, MediaWiki* and other development-related namespaces, as suggested by the large amount of namespaces, users assigned to this user role contribute to a diversity of technique-related namespaces and hence the name for this user role; user role "Society Technology" and "Nature Science" mainly relate to content contribution to society, technology, nature and science-related Wikipedia articles. Note that different user roles correspond to different common work archetypes.

Figure 2 presents the trend of user roles over time. We observe that "Society Technology" is the most dominant user role with about 25 % of user profiles being assigned to it; other dominant user roles include "Politics", "People", "Arts" and "Maintaince"; each of the remaining user roles related to the maintenance,

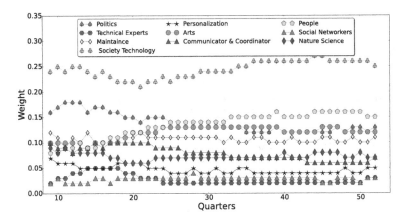

Fig. 2. Popularity of common user roles over time. *Quarters* corresponds to the index of quarter, *Weight* indicates the weight of user roles in each quarter, where the weight is calculated as the percentage of user profiles assigned to that user role in a quarter. Each curve represents the trend of each user role over time.

management and organization aspects of Wikipedia generally accounts for less than 10 % of user profiles. One obvious trend in Fig. 2 is that the majority of user profiles are assigned to user roles related to content contribution to Wikipedia articles; where a substantial amount of user profiles are assigned to maintenance related user roles.

Community-Level Change in User Roles. As users stay long enough in Wikipedia, they tend to shift their participation by changing their edits to other namespaces and categories of articles. Moreover, the addition of new functionalities is also likely to encourage shift in user participation. Figure 3 visualizes the evolution of three user roles at an aggregate level, from which we make several observations. Firstly, in Fig. 3(a), the Wikipedia namesspace is the most dominant features in user role "Technical Experts" with weight larger than 0.65, followed by the Wikipedia Talk namespace, where other features generally have weights less than 0.01. By contrast, in Fig. 3(b), we observe that the weights are distributed more equally among the features and the weights of features change gradually over time. For instance, the weight of the Society category increases gradually from Quarter 10 to 30 and then become relatively stable onwards; the weight of the Mathematics category decreases gradually from Quarter 10 to 20 and stays around 0.03 from Quarter 20 onwards. Lastly, in Fig. 3(c), user role "Arts" is dominated by two categories of articles: Arts (with a gradual increasing trend) and Humanities (with a slow decreasing trend).

The other user roles evolve in similar ways as those in Fig. 3, but are omitted here. The evolution of user roles captures the overall changes in user profiles due to the addition of new functionalities and shifts in user participation over time, but it provides little insight about what is the general trajectory of a user's participation as the user transitions from being a newcomer to being an

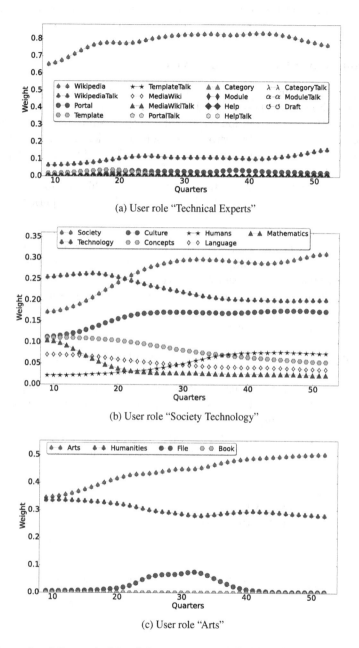

(a) User role "Technical Experts"

(b) User role "Society Technology"

(c) User role "Arts"

Fig. 3. Example of the evolution of the common user roles. *Quarters* corresponds to the index of quarter, *Weight* indicates the weight of features for the user roles, where the weight of each feature corresponds to the importance of the feature in the DTM topic. Each curve represents the evolution of the weight of a feature over time.

established member of the community. We will explore this question in the next section.

4.2   Group Level Change in Lifecycle

Different from previous studies which analyze user lifecycle on a medium size dataset (with about 33,000 users [7,8,13]), our analysis is based on a large dataset with approximately half a million users. Different users are more likely to follow slightly or totally different trajectories in their lifecycle. As a result, it is very unlikely for us to employ one single model to perfectly fit the development of lifecycle for all users. For instance, Rowe [13] used linear regression model to characterize the development of user properties. In this section, we examine how different groups of users evolve throughout their lifecycle periods from two perspectives:

- Analyzing the variation within the distribution of editor profiles toward user roles in each lifestage period, thus informing how user changes edit preference among namespaces and categories of articles.
- Analyzing the change in variation when comparing each user's profile in one period with previous periods, thus indicating how the user changes edit activity relative to past behaviour.

In this analysis, we are interested in exploring questions such as: what are the differences between the changes in edit activity for different groups of users (e.g. short-term vs. long-term users, admin vs. bot[9] users)? For this purpose, we divide the users into several groups according to the number of active quarters as in Table 3. Understanding how individual users develop over time in online communities makes it possible to develop techniques for important tasks such as churn prediction, as we will show through experiments in Sect. 5.

Table 3. Groups of editors with different number of active quarters, which are further divided into three categories: *Hyper active*, *General groups* and *Hibernate*.

Name	*Hyper active*			*General groups*		*Hibernate*
	≥40	30–35	Admins	Bots	All	4
#Active quarters	more than 40	30 to 35	(previous or current) administrators	bot users	All users	equal to 4

Comparison of Individual Periods (Period Entropy). Entropy is often used to describe the amount of variation in a random variable, and provides

---

[9] In Wikipedia, bots are generally programs or scripts that make repetitive automated or semi-automated edits without the necessity of human decision-making: http://en.wikipedia.org/wiki/Wikipedia:Bot_policy.

a powerful tool to quantify how much a specific user changes his edit activity within each lifecycle period. Following [13], the entropy of an arbitrary probability distribution P is defined as follows:

$$H(P) = -\sum_x p(x)\log p(x) \tag{3}$$

We calculated the entropy of each user throughout their lifestages based on the distributions of their profiles toward the 10 common user roles, and then recorded the mean of the entropy measures for each group within each period. Plotting the mean entropy values over lifecycle periods provides an overview of the general changes that each group experiences. Figure 4(a) visualizes the period entropies for different groups of editors over time.

Contrasts of Historical Periods (Cross-Period Entropy). While period entropy captures how a user changed edit preference over time, it neglects how a user changes edit preference compared to its previous edit preference. Cross-period entropy overcomes this limit and can be used to gauge the cross-period variation in user's edit activity throughout lifecycle periods. The cross-entropy of one probability distribution P (from a given lifecycle period) with respect to another distribution Q from an earlier period (e.g. the previous quarter) is defined as follows [13]:

$$H(P,Q) = -\sum_x p(x)\log q(x) \tag{4}$$

We computed the cross-period entropy for each user group in the same manner as period entropy. Figure 4(b) presents the cross-entropies for different groups across lifecycle periods. We make the following observations from the evolution of entropies in Fig. 4:

– There is an observable seperation between the period entropies and cross-entropies of different groups (particularly for the *Hyper active* and *Hibernate* categories), suggesting an obvious difference exists in the period and cross-period variation in edit activity for different groups of users.
– The period entropies of groups with long-term users (i.e. $\geq$40, 30–35 and Admins) are much higher than those of groups with short-term users (i.e. 4), indicating that long-term users generally have more diversified edit preference than those short-term users. The period entropies of groups with long-term users (i.e. the *Hyper active* category) present a slow increase in the early stage of lifespans followed by a gradual decrease in the late stage of lifespans, suggesting that long-term users gradually diversify their participation by shifting edit preference and then their edit preference become stable as they spend more time in the community.
– On the other hand, the cross-entropies of groups with short-term users (i.e. 4) are much higher than those of groups with long-term users (i.e. $\geq$40, 30–35 and Admins), suggesting that short-term users generally experience more fluctuation in their historical edit activity and thus distribute their edit contribution at random among multiple namespaces and categories of articles

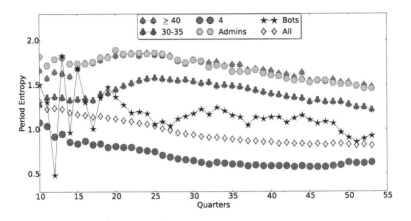

(a) Period entropies

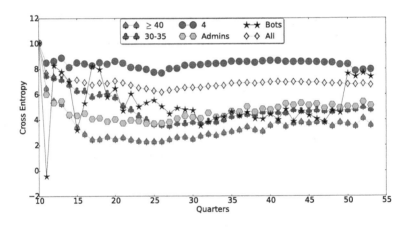

(b) Cross-period entropies

Fig. 4. Evolution of (a) period entropies derived from the distributions of user profiles toward the 10 common user roles, (b) cross-period entropies formed from comparing user profile distribution with previous lifestage periods.

over the course of their career. The cross-entropies of all groups experience a decrease in the early stage of lifecycle followed by a stable change onwards, suggesting that comparing with historical behaviour, all users are more likely to change edit preference in the beginning than in the late stage of their lifecycle.

To further validate the reasonableness of the above groupings and the significance of the observations, we perform a multiple comparison of group means using Tukey's honest significant difference (HSD) test[10] [14]. It is a popular test with good statistical properties for the pairwise comparison of means with k samples. The null hypothesis H_0 of Tukey's HSD test being: the two samples have the same mean, i.e. $\mu_1 = \mu_2$. Table 4 gives the statistics of rejection in the Tukey's HSD test for the pairwise comparison of means with the 6 groups given in Table 3 over 44 quarters (i.e. from Quarter 10 to Quarter 53).

Table 4. Statistics of rejection for Tukey's HSD test (significant level: $P < 0.05$). The column *Reject H_0* refers to the number of quarters in which the means of the pairwise groups are significantly different. H_0: The two groups have the same mean, i.e. $\mu_1 = \mu_2$.

Pairwise groups		Period entropy/Cross-period entropy	
Group1	Group2	*Reject H_0* (i.e. $\mu_1 \neq \mu_2$)	*Cannot reject H_0* (i.e. $\mu_1 = \mu_2$)
≥ 40	30–35	**44**	0
≥ 40	4	**42**	2
≥ 40	Admins	3	**41**
≥ 40	Bots	**36**	8
≥ 40	All	**44**	0
30–35	4	**41**	3
30–35	Admins	**44**	0
30–35	Bots	**32**	12
30–35	All	**39**	5
4	Admins	**43**	1
4	Bots	**37**	7
4	All	**40**	4
Admins	Bots	**38**	6
Admins	All	**44**	0
Bots	All	21	23

A very obvious trend in Table 4 is that for both period entropy and cross-period entropy, the means of the *Hyper active* category (i.e. groups ≥ 40, 30–35 and Admins) are significantly different from those of the *General groups* and *Hibernate* categories in the majority of quarters. For instance, the mean of group ≥ 40 is significantly different with that of the whole population (i.e. group All) in all the 44 quarters, is significantly different with that of the *Hibernate* category in 42 out of 44 quarters, but is not significantly different from that of group Admins in 41/44 quarters. It is interesting to observe that the mean of

---

[10] The implementation of the test for R and *Python* enviroment can refer to: http://jpktd.blogspot.ie/2013/03/multiple-comparison-and-tukey-hsd-or_25.html.

group Bots is significantly different from that of the *Hibernate* category in 37/44 quarters. These results provide further supports for the observations from Fig. 4.

To summarize, the above observations and significant test suggest that: long-term editors generally have more diversified edit preference and less cross-period variation in their edit activity than their short-term counterparts; by contrast, short term editors generally do not develop long term edit interest and tend to distribute their edit contribution at random among multiple namespaces and categories of articles, thus experience more fluctuation in their profile distributions. The observations suggest that the fluctuation in profile assignments can be used as one of the indicators to identify users' exist in the community. In the next section, we will generate features corresponding to these findings for the churn prediction task.

4.3   User Level Change in Lifecycle

In this section, we provide sample individual user lifecycles for three groups of editors: bot, short-term and long-term users (including admin users) in terms of the evolution of profile distribution over time. These user lifecycles provide further supports for the arguments we make in the previous section. Figure 5 presents the lifecycle of selected users, which provides a direct and obvious support for the observations from the evolution of entropies in Fig. 4:

- There are certain level of fluctuations in the profile distributions of short-term users throughout lifecycle periods (Fig. 5(a), (b)), signifying that short-term users generally do not develop long-term edit interest and tend to distribute their edit contribution at random among multiple namespaces and categories of articles over the course of their career in the community.
- One the other hand, long-term users generally experience gradual and soft evolution in their profile distributions, and have more diversified edit preference than short-term users; these users tend to have focused/dominant long-term edit interest in one or more namespaces/categories of articles throughout their contributory lifespans (Fig. 5(d), (e), (f)).
- By constract, the profile distributions for bot users are slightly complex and need special care for explanation. There are three categories of profile distributions for bot users: (1) similar profile distributions as those short-term users (Fig. 6(a)); (2) almost constant profile distributions throughout the lifecycle (Fig. 6(b)), indicating bots that perform single task in Wikipedia; (3) profile distributions that are very similar to those long-term users (Fig. 6(c), (d)).

Overall, the plots of individual user lifecycles suggest that there are different patterns of change in profile distributions for different categories of editors, and that the features inspired by these changes might be beneficial for practical tasks such as churn prediction.

5   Application of Editor Lifecycles to Churn Prediction

We have so far focused on understanding the lifecycle trajectories of editors based on their exhibited behaviour. We now move on to exploring how predictive are

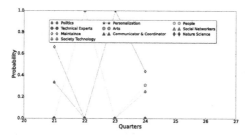

(a) Short-term user stayed for 4 quarters

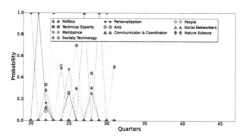

(b) Short-term user stayed for 10 quarters

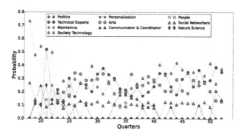

(c) Long-term user (Admin)

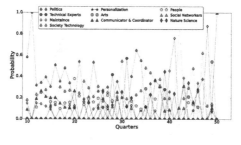

(d) Long-term user (Admin)

Fig. 5. Examples of the dynamic of profile distribution for selected short-term and long-term users. *Probability* indicates the probability of assigning user profile to a specific user role, where the probabilities were predicted by DTM.

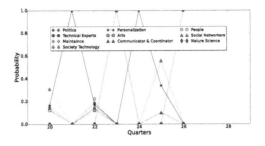

(a) Short-term bot user 2

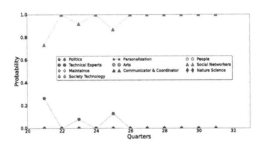

(b) Short-term bot user 1

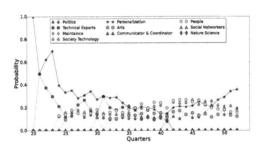

(c) Long-term bot (DumbBOT)

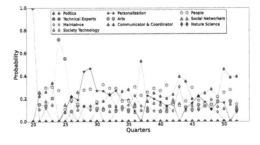

(d) Bot account operated by Admin (Cydebot)

Fig. 6. Examples of the dynamic of profile distribution for selected bot accounts.

the features generated from patterns of change in editor profiles in identifying whether a user will abandon a community. Churners present a great challenge for community management and maintenance as the turnover of established members can have a detrimental effect on the community in terms of creating communication gap, knowledge or other gaps.

5.1 Definition of Churn Prediction

Following the work by Danescu-Niculescu-Mizil *et al.* [7], we define the churn prediction task as predicting whether an editor is among the 'departed' or the 'staying' class. Considering that our dataset spans more than 13 years (i.e. 53 quarters), and that studies about churn prediction generally follow the diagram of predicting the churn status of users in the prediction period based on user exhibited behaviour in the observation period (e.g. [7,16]), we employed a sliding window based method for churn prediction. Specifically, we make predictions based on features generated from editor profile distributions in a sliding window with $w=4$ quarters. An editor is in the 'departed' class if she left the community before active for less than $m=1$ quarter after the sliding window, denote $[w, w + m]$ as the departed range. Similarly, an editor is in the 'staying' class if she was active in the community long enough for a relatively large number of quarters ($n \geq 3$) after the sliding window, term $[w + n, +\infty]$ as the staying range.

5.2 Features for the Task

Our features are generated based on the findings reported in the previous section. For simplicity, we assume the w quarters included in the i-th sliding window being $i = [j, \cdots, j + w - 1]$ ($j \in [10, 50]$), and denote the Probability Of Activity Profile of an editor in quarter j assigned to the k-th user role as $POAP_{i,j,k}$. We use the following features to characterize the patterns of change in editor profile distributions:

- *First active quarter*: the quarter in which an editor began edits in Wikipedia. The timestamp a user joined the community may affect her decision about whether to stay for longer time as previous study suggested that users joined later in Wikipedia may face a more severe situation in terms of the likelihood of contribution being accepted by the community and the possibility of being excluded by established members.
- *Cumulative active quarters*: the total number of quarters an editor had been active in the community till the last quarter in the sliding window.
- *Fraction of active quarters in lifespan*: the proportion of quarters a user was active till the sliding window. For instance, if an editor joined in quarter 10, and stayed active for 8 quarters till the current sliding window (16–19), then the figure is calculated as: $8/(19-10+1) = 0.80$.
- *Fraction of active quarters in sliding window*: the fraction of quarters a user was active in current sliding window.

- *Similarity of profile distribution in sliding window*: quantifies the similarity of user profile distributions in any two successive quarters using cosine similarity.
- *Diversity of edit activity*: denotes the entropy of $POAP_{i,j,k}$ for each quarter j in window i, calculated using Eq. (3). This measure captures the extent to which an editor diversified her edits toward multiple namespaces and categories of articles.
- *Cross-entropy of edit activity*: denotes the historical variation in $POAP_{i,j,k}$ compared to the same measure in previous quarters, calculated using Eq. (4). This measure captures the extent to which an editor changed her edit activity compared to her past behaviour.
- *mean* $POAP_{i,j,k}$: denotes the average of $POAP_{i,j,k}$ for each user role k in window i. This measure captures whether an editor focused her edits on certain namespaces and categories of articles in window i.
- $\Delta POAP_{i,j,k}$: denotes the change in $POAP_{i,j,k}$ between the quarter $j - 1$ and j (for $j \in [2, 50]$), measured by $\Delta POAP_{i,j,k}=(POAP_{i,j,k} - POAP_{i,j-1,k} + \delta)/(POAP_{i,j-1,k} + \delta)$, where δ is a small positive real number (i.e. 0.001) to avoid the case when $POAP_{i,j-1,k}$ is 0. This measure also captures the fluctuation of $POAP_{i,j,k}$ for each user role k in window i.

For each editor, the first three features are global-level features which may be updated within the sliding window, the remaining features are window-level features and are recalculated within each sliding window. The intuition behind the last four features is to approximate the evolution of editor lifecycle we sought to characterize in the previous section. The dataset is of the following form: $D = (x_i, y_i)$, where y_i denotes the churn status of the editor i, $y_i \in \{\text{Churner, Non-churner}\}$; x_i denotes the feature vector for i.

5.3   Performance of Churn Prediction

Experimental Setup. The prediction task is a binary classification problem, we use the logistic regression model for the purpose. To avoid possible bias in the results due to the imbalance of class distribution, for each sliding window, we randomly sampled the editors in order to generate a dataset with the desired class ratio (churners:non-churners) being 1:2. Each time we moved the window by one quarter. The results reported in the following are averaged over 10-fold cross-validation.

Performance of Sliding Window based Churn Prediction. Figure 7 plots the performance measures of churn prediction. We observe that the performance measures are relatively stable across different windows: **ROC area** increased from 0.75 to **0.80**; **FP rate** decreased from 0.40 to 0.34; other measures (i.e. **Precision, Recall, F-measure**) stayed around **0.74**. Note when Danescu-Niculescu-Mizil *et al.* [7] performed churn prediction on two beer rating communities (i.e. BeerAdvocate and RateBeer) based on user's linguistic change features, they obtained the best performance of **Precision** being 0.77, **Recall** being 46.9, **F-measure** being 0.56. Rowe [13] evaluated the churn prediction task on three online platforms (i.e. Facebook, SAP and Server Fault) using user's

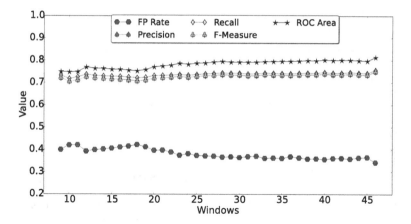

Fig. 7. Performance for sliding window based churn prediction. *Windows* represents the sliding window, *Value* represents the performance.

social and lexical change related features, and obtained the best performance of **Precision@K** being 0.791 and **AUC** being 0.617. Comparing with the statistics from [7,13], the results suggest that although our study makes use of user activity related features different from those in [7,13], our approach achieves at least comparable overall and average performance with those two studies for churn prediction in online communities. This observation suggests that in online communities, sudden change in user activity can be an important signal that the user is likely to abandon the community.

Cumulative Gains for Churn Prediction. Lift factors are widely used by researchers to evaluate the performance of churn-prediction models (e.g. [16]). The lift factors achieved by our model are shown in Fig. 8. In a lift chart, the diagonal line represents a **baseline** which randomly selects a subset of editors as potential churners, i.e., it selects s% of the editors that will contain s% of the true churners, resulting in a lift factor of 1. In Fig. 8, on average, our model with *all features* was capable of identifying 10 % of editors that contained 21.5 % of true churners (i.e. a lift factor of 2.15), 20 % of editors that contained 40 % of true churners (i.e. a lift factor of 1.999), and 30 % of editors that contained 55.5 % of true churners (i.e. a lift factor of 1.849). The model with only *Global active quarter stats* was able to identify 10 % of editors that contained 19 % of true churners (a lift factor of 1.903), 20 % of editors that contained 33.7 % of true churners (a lift factor of 1.684), and 30 % of editors that contained 48.1 % of true churners (a lift factor of 1.604). Evidently, all our models achieved higher lift factors than the baseline, and the model with *all features* obtained the best lift chart. Thus if the objective of the lift analysis is to identify a small subset of likely churners for an intervention that might persuade them not to churn, then this analysis suggests that our model can identify a set of 10 % of users where the probability of churning is more than twice the baseline figure.

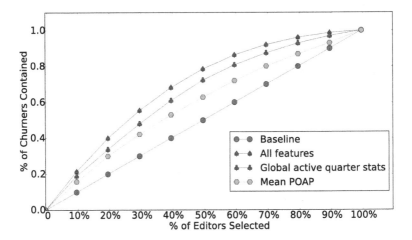

Fig. 8. Lift chart obtained by the proposed churn-prediction model. The statistics presented are averaged over all sliding windows. Different curves represent the lift curves for different groups of features.

Table 5. Performance of feature analysis. Bold face numbers indicate best performing features for the corresponding metric. The performance metrics reported are the average values over all sliding windows.

Performance improvement by incrementally adding different feature sets to the base model					
Feature set	FP rate	Precision	Recall	F-Measure	ROC Area
Global active quarter stats	0.45 ± 0.08	0.69 ± 0.02	0.70 ± 0.01	0.68 ± 0.04	0.72 ± 0.04
+ Similarity of profile distribution	0.42 ± 0.03	0.70 ± 0.01	0.72 ± 0.01	0.70 ± 0.01	0.75 ± 0.02
+ Diversity of edit activity	**0.38** ± 0.02	**0.73** ± 0.01	**0.74** ± 0.01	**0.73** ± 0.01	**0.78** ± 0.02
+ Cross-entropy of edit activity	0.38 ± 0.02	0.73 ± 0.01	0.74 ± 0.01	0.73 ± 0.01	0.78 ± 0.02
+ mean POAP$_{i,j,k}$	0.38 ± 0.02	0.73 ± 0.01	0.74 ± 0.01	0.73 ± 0.01	0.78 ± 0.02
+ ΔPOAP$_{i,j,k}$	0.38 ± 0.02	0.73 ± 0.01	0.74 ± 0.01	0.73 ± 0.01	0.78 ± 0.02
Performance for individual feature sets					
All features	**0.38** ± 0.02	**0.73** ± 0.01	**0.74** ± 0.01	**0.73** ± 0.01	**0.78** ± 0.02
Global active quarter stats	0.45 ± 0.08	0.69 ± 0.02	0.70 ± 0.02	0.68 ± 0.04	0.72 ± 0.04
Similarity of profile distribution	0.50 ± 0.02	0.66 ± 0.01	0.69 ± 0.01	0.66 ± 0.01	0.70 ± 0.01
Diversity of edit activity	0.43 ± 0.01	0.70 ± 0.01	0.71 ± 0.01	0.70 ± 0.01	0.75 ± 0.01
Cross-entropy of edit activity	0.66 ± 0.01	0.49 ± 0.08	0.67 ± 0.003	0.54 ± 0.01	0.57 ± 0.01
mean POAP$_{i,j,k}$	0.60 ± 0.01	0.64 ± 0.01	0.67 ± 0.004	0.60 ± 0.01	0.63 ± 0.01
ΔPOAP$_{i,j,k}$	0.60 ± 0.02	0.63 ± 0.01	0.67 ± 0.004	0.60 ± 0.016	0.64 ± 0.02

Feature Analysis. To better understand the contribution of each feature set in the context of all the other features, we performed an ablation study. Specifically, we added the feature sets incrementally to the base model containing only *Global active quarter stats* features, and ran a logistic model on the incremental dataset. We then compared the performance of the resulting classifier to that of classifier with the complete set of features. We observe that the first three feature sets bring substantial improvements in performance when added incrementally to the base model (Table 5); the last three feature sets bring very little improvements in performance when added incrementally to the model. Performance for individual feature sets provides further support for the observation: for indivitual feature sets, *Diversity of edit activity* is the most discriminative feature set with the best performance for the task, *Global active quarter stats* is the second best, *Similarity of profile distribution* is the third best; *Cross-entropy of edit activity* is the least discriminative feature set for the task. We also observe that none of the feature sets can by itself achieve the performance reported in Fig. 7, which suggests that the feature sets we generated (at least the first three feature sets) complement each other in predicting churners in online communities. The results suggest that sudden change in user behaviour can be a signal that the user is prone to churning from the community.

6   Conclusion

In this work we have presented a novel latent space analysis of editor lifecycles in online communities with preliminary results. The latent space representation reveals a number of different categories of editor (e.g. *Technical Experts, Social Networkers*) and provides a means to track the evolution of editor behaviour over time. We also perform an entropy-based analysis of editor profiles in the latent space representation and find that long term and short term users generally have very different profile distributions and evolve differently in their lifespans. The presented approach for understanding online user behaviour is a novel one in terms of its generality and interpretability. The approach can be applied to model time-varying user behaviour data whenever the data are available, with intuitive and interpretable results.

We show that understanding patterns of change in user behaviour can be of practical importance for community management and maintenance, in that the features inspired by our latent space analysis can differentiate churners from non-churners with reasonable performance. This work opens interesting questions for future research. Firstly, social networking plays a foundational role in online communities for knowledge sharing and member retention, future work should accommodate social dynamics in the model. Combining user activity features with social network and/or lexical features in a unified framework would allow us to explore interesting questions, such as how individual users grow to become experienced/established members in the community by interacting or social learning with other senior members. Secondly, the topic analysis we used in this paper did not account for the evolution of activity in user-level, it will be very helpful to develop topic models that accommodate the evolution in the user and community level. Moreover, online communities are very similar in terms of allowing multi-dimensions of user activity, the approach presented in this study can be generalized to other communities very easily.

Acknowledgements. This work is supported by Science Foundation Ireland (SFI) under Grant No. SFI/12/RC/2289 (Insight Centre for Data Analytics). Xiangju Qin is funded by University College Dublin and China Scholarship Council (UCD-CSC Joint Scholarship 2011).

References

1. Ahmed, A., Low, Y., Aly, M., Josifovski, V., Smola, A.J.: Scalable distributed inference of dynamic user interests for behavioral targeting. In: Proceedings of KDD, pp. 114–122. ACM (2011)
2. Ahmed, A., Xing, E.P.: Timeline: a dynamic hierarchical dirichlet process model for recovering birth/death and evolution of topics in text stream. In: Proceedings of UAI, pp. 20–29 (2010)
3. Blei, D.M., Lafferty, J.D.: Dynamic topic models. In: Proceedings of ICML, pp. 113–120 (2006)
4. Blei, D.M., Ng, A.Y., Jordan, M.I.: Latent dirichlet allocation. J. Mach. Learn. Res. **3**, 993–1022 (2003)
5. Boutsidis, C., Gallopoulos, E.: SVD based initialization: a head start for non-negative matrix factorization. In: Pattern Recognition (2008)
6. Chan, J., Hayes, C., Daly, E.M.: Decomposing discussion forums using user roles. In: Proceedings of ICWSM, pp. 215–218 (2010)
7. Danescu-Niculescu-Mizil, C., West, R., Jurafsky, D., Leskovec, J., Potts, C.: No country for old members: user lifecycle and linguistic change in online communities. In: Proceedings of WWW, pp. 307–318. Rio de Janeiro, Brazil (2013)
8. Furtado, A., Andrade, N., Oliveira, N., Brasileiro, F.: Contributor profiles, their dynamics, and their importance in five Q&A sites. In: Proceedings of CSCW, pp. 1237–1252, Texas (2013)
9. Hulpus, I., Hayes, C., Karnstedt, M., Greene, D.: Unsupervised graph-based topic labelling using dbpedia. In: Proceedings of WSDM, pp. 465–474. ACM (2013)
10. Lin, C.: Projected gradient methods for non-negative matrix factorization. Neural Comput. **19**(10), 2756–2779 (2007)
11. O'Callaghan, D., Greene, D., Carthy, J., Cunningham, P.: An analysis of the coherence of descriptors in topic modeling. Expert Syst. Appl. **42**(13), 5645–5657 (2015)
12. Panciera, K., Halfaker, A., Terveen, L.: Wikipedians are born, not made: a study of power editors on wikipedia. In: Proceedings of GROUP, pp. 51–60. ACM (2009)
13. Jin, Y., Zhang, S., Zhao, Y., Chen, H., Sun, J., Zhang, Y., Chen, C.: Mining and information integration practice for chinese bibliographic database of life sciences. In: Perner, P. (ed.) ICDM 2013. LNCS, vol. 7987, pp. 1–10. Springer, Heidelberg (2013)
14. Tukey, J.W.: Comparing individual means in the analysis of variance. Biometrics **5**(2), 99–114 (1949)
15. Wang, X., McCallum, A.: Topics over time: a non-Markov continuous-time model of topical trends. In: Proceedings of KDD, pp. 424–433. ACM (2006)

16. Weia, C.P., Chiub, I.T.: Turning telecommunications call details to churn prediction: a data mining approach. Expert Syst. Appl. **23**(2), 103–112 (2002)
17. Welser, H.T., Cosley, D., Kossinets, G., Lin, A., Dokshin, F., Gay, G., Smith, M.: Finding social roles in wikipedia. In: Proceedings of iConference, pp. 122–129. ACM (2011)

On Spatial Measures of Geographic Relevance for Geotagged Social Media Content

Xin Wang, Tristan Gaugel[✉], and Matthias Keller

Steinbuch Centre for Computing, Institute of Telematics,
Karlsruhe Institute of Technology, 76131 Karlsruhe, Germany
uidhw@student.kit.edu, {tristan.gaugel,matthias.keller}@kit.edu

Abstract. Recently, geotagged social media contents became increasingly available to researchers and were subject to more and more studies. Different spatial measures such as Focus, Entropy and Spread have been applied to describe geospatial characteristics of social media contents. In this paper, we draw the attention to the fact that these popular measures do not necessarily show the geographic relevance or dependence of social content, but mix up geographic relevance, the distribution of the user population, and sample size. Therefore, results based on these measures cannot be interpreted as geographic effects alone. By means of an assessment, based on Twitter data collected over a time span of six weeks, we highlight potential misinterpretations and we furthermore propose normalized measures which show less dependency on the underlying user population and are able to mitigate the effect of outliers.

1 Introduction

Although today's technical infrastructure allows communication with almost anybody, anywhere in the world, geographic location plays a role in the selection of communication partners and topics. Geospatial aspects of online communication have attracted more and more interest from the research community: Which role does location play in the selection of online friends (e.g. [10])? How strong is the local focus of online social media contents and which geospatial spreading patterns can be observed (e.g., [2,4])? These and other questions have been examined to gain insights into the mechanisms of social media usage and content dispersion. Social media have increasing impact on the spreading of news [9], but in contrast to traditional newswire, the mechanisms of distribution are complex and rarely understood yet.

In this paper, we demonstrate that popular measures, such as Spread, Entropy, and Focus do not necessarily only show the geographic relevance or dependence of social content, but are severely influenced by the distribution of the user population, and sample size. Therefore, results based on these measures cannot be interpreted as geographic effects alone. Furthermore, we propose normalized measures which that show less dependency on user population. For the remainder of this paper, we assume that we are dealing with geographic coordinates but abstract from the way they have been retrieved. A set of coordinates

© Springer International Publishing Switzerland 2016
M. Atzmueller et al. (Eds.): MSM, MUSE, SenseML 2014, LNAI 9546, pp. 70–89, 2016.
DOI: 10.1007/978-3-319-29009-6_4

can represent, for instance, the occurrences of a hashtag. For readability, we will call such a set of coordinates the occurrences of a meme, although it could represent, for example, the locations of the friends of a Facebook user as well. Many studies are not based on the analysis of a single meme but on the analysis of a large set of memes, so that all individual sets of meme occurrences cannot be visualized or manually inspected. Hence, measures are required that capture geographic properties of the memes, e.g., to find the hashtags that have the strongest local concentration.

Based on such kind of measures, recent studies revealed surprising patterns of geo-spatial distribution: For instance, Kamath et al. [4] found in a study of geotagged tweets that hashtags seem to be most concentrated at a single location at their peak time, reaching the point of lowest concentration in average after 20 min, to slowly return to a single location afterwards. In an analysis of YouTube videos [2] a very similar distribution pattern was observed, called spray-and-diffuse-pattern by the authors. Kamath et al. [4] suggest that this is a fundamental pattern of social media spread. Can we conclude that the users' interest in a meme is traveling, e.g. that at the peak time, a meme is relevant only to users in a single location while it gets increasingly interesting for distant users within the next 20 min?

In order to answer such questions, one should make sure that the employed measures are able to capture these aspects and are not severely influenced by other effects, such as, e.g., the spatial distribution of the user population. Studying this characteristic and drawing conclusions is particularly challenging, since multiple conditions have to be met in order to be able to observe a meme occurrence at coordinates (x, y): (a) a user was present at position (x, y), (b) due to its relevance the user shared the meme, and (c) location information for this meme is available. The study of geospatial properties of memes is based on the hypothesis that the relevance of a meme to a user depends on his location (x, y). In this paper, we consider the geographic relevance of a meme as the probability that a user contributes to the meme in dependence of his location (x, y). The geographic relevance represents the user behavior. Hence, if we are interested in drawing conclusions about the user behavior, we should focus on the geographic relevance (b) and try to abstract from the user population distribution (a) and the sample size (c).

We assume a basic, circular model for the geographic relevance: only users within a circular area of radius r contribute to a specific meme with a probability $p > 0$. If the radius r is small, the users' interest is concentrated and if it is large, the user's interest in the meme is more widespread. This model is surely a simplification, but it allows to examine how well traditional geospatial measures reflect the geographic relevance and to develop more suitable measures for scenarios with focus on geographic relevance.

The relationship between the concentration of the geographic relevance represented by the radius r and the traditional measures is often counterintuitive as we exemplify with Fig. 1: A traditional measure applied in geographic statistics that was also used to analyze social media data (e.g. [4]) is the mean spread, the average distance of samples to their geographic mean point. Mean spread is interpreted as measure for the geographic dispersion of a set of meme occurrences. In Fig. 1-A, we assume that a specific topic, identified by a particular meme, has a strong local focus in central France. We modeled this by choosing a center point, and assume that r=150 km. Furthermore, we assume that 50 geotagged occurrences of memes (e.g. with the same particular hashtag) appear within the focal radius. To model the distribution of user population, we collected real geotagged Tweets from this area and selected a random subset of 50 samples. Based on this sample, we calculated a mean spread of 90.6 km. Now, we assume that the meme gets interesting to users in a larger area and set r=250 km, generate a sample in exactly the same way and calculate the mean spread again (Fig. 1-B). However, the measured mean spread does not increase but drop to 66.2 km. This is caused by the fact that Paris is now within the circular area and the topic has spread to an area in which the population is much more concentrated.

In this paper, we analyze the impact of the users' distribution and the potentially limited number of available samples on the way in which the measures reflect the geographic relevance of memes. The structure and the contributions of this paper are as follows: In Sect. 2, we review related work and introduce three common measures: Focus, Spread and Entropy. Section 3 then formalizes the problems of the influence of (i) the user distribution and (ii) too small sample sizes in order to specify corresponding requirements for the measures. Furthermore, we introduce our methodology of assessing the impact of these problems in realistic scenarios with Monte Carlo simulations. In Sect. 4, we discuss the simulation results and show that both, the user distribution and the sample size severely limit the meaningfulness of geospatial measures with respect to geographic relevance and show that a relatively high number of samples is necessary to achieve indicative results. In Sect. 5, we propose and evaluate modified measures that are more robust against population density effects. Furthermore, we explain, why it is not possible to correct for systematic errors of Focus and Entropy caused by a varying numbers of samples. Section 6 concludes the paper

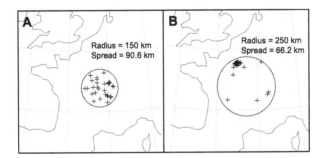

Fig. 1. Influence of user population on spatial measures: Although the focus area is larger in plot B, a smaller Spread is measured.

and we summarize the implications for existing studies and the application of the measures in the future.

2   Related Work

In Sect. 2.1, we discuss related work. In Sect. 2.2 we specify the three common spatial measures Focus, Entropy and Spread formally.

2.1   Geospatial Analysis of Social Media Contents

Due to the recent increasing interest in analyzing geospatial properties of social media, plenty of measures have been employed by the research community in order to quantify different aspects of geo-spatiality. The variety of measures also emerges from the fact that the underlying data varies in its characteristics. While for some works [4,7] the analyzed geospatial data originates from Twitter, other studies employed similar analyses for YouTube videos [2] or web resources [3]. The specific characteristics required slightly varying definitions for similar measures. The three measures which are considered and analyzed in this work, namely the Spread, Entropy, and Focus have all been used in a slightly adjusted way in several works before.

Spread is used as a measure for uniformity and in general states the mean distance over all geo-locations to the geographical midpoint. It is for example being used in [4] for an analysis of the dispersion of Twitter users or in [3] for analyzing geographical scopes of web resources. Entropy is generally used in accordance with its information theoretical meaning of stating the bits required to represent the spatial spread. The granularity of the spatial resolution as well as the actual spatial distribution of, e.g., the users or tweets thus plays an important factor in determining the entropy. A high value is seen as an indication for a more uniform spread of the data under consideration. Entropy is hereby employed by [2,4] as an indication for the randomness in spatial distribution for tweets and YouTube video views, respectively. Focus is used as a measure for the centricity of tweets [4] and YouTube video views [2] by calculating the maximum proportion of occurrences of geo-locations falling into any individual area.

Additionally, there are other works that focus less on calculating a single measure for representing the spatiality, but represent it, e.g., by means of a CDF of the geographic distance between users of an social network [10]. In [7] again a density measure is used to visualize areas with high or low occurrences of tweets. Then again other work primarily focuses on identifying the center point and the corresponding dispersion for information that is spatially distributed, e.g. search queries [1] or web resources [3]. Although these studies are dealing with a similar topic of quantifying spatiality, their goal is not to represent this by means of one single measure.

2.2   Spatial Measures

In this paper, we analyze the measures Focus, Entropy and Spread, which were used in previous work. Focus and Entropy are zone-based measures that require the globe to be divided into a set R of zones (e.g., countries [2] or identical squares of 10 Km by 10 km [4]). If the set O denotes the set of all meme occurrences and $O_i \in O$ denotes the occurrences in zone $i \in R$, the relative frequency P_i of zone i is given by:

$$P_i = \frac{|O_i|}{|O|} \tag{1}$$

Focus measures whether there is a single zone in which the meme occurrences are concentrated. Focus is defined as the maximum P_i for any region i. Then, i is called focus location. As a consequence, the higher the focus value is, the more occurrences appear in the focus location and the fewer in other regions.

$$Focus, F = \max_{i \in R} P_i \tag{2}$$

Entropy measures whether the occurrences are concentrated in only a few zones (low entropy) or whether the samples are more equally distributed over a large amount of zones (high entropy).

$$Entropy, E = -\sum_{i \in R} P_i log_2 P_i \tag{3}$$

Spread measures how close the occurrences are located to each other, without the usage of a zone system. Spread is defined as the mean distance of all meme occurrences to their geographical mean point. Higher values of Spread correspond to a larger spatial coverage of the distribution. With $D(c_1, c_2)$ being the geographic distance between two points c_1 and c_2, Spread is defined as:

$$Spread, S = \frac{\sum_{o \in O} D(o, MeanPoint)}{|O|} \tag{4}$$

where the *MeanPoint* equals the geometric median, the location with the smallest possible average distance to all $o \in O$.

3   Methodology

In this section, we first formalize the requirements for spatial measures that reflect the geographic relevance (Sect. 3.1). In order to analyze the impact of a non-uniform user population, we collected an extensive amount of geo-tagged tweets from Twitter which we are using as an approximate of the user population U (cf. Sect. 3.2). Since considering the impact of this real-world data set would be hard to realize with an analytical approach, we conduct Monte Carlo simulations (Sect. 3.3). Challenges related to partitioning are explained in Sect. 3.4 since both, the Focus and Entropy are zone-based measures.

3.1   Formalization of Requirements

For the remainder of this paper we use the term *geographic relevance* of a meme as the probability that a user publishes a meme within a certain time span in dependency of the user location. We assume that this probability can be quantified by $f(x, y)$, expressing the probability for a user to contribute to a meme, if located at (x, y). Consequently, $f(x, y)$ models the geographic relevance of a meme. While we assume that such a geographic relevance function exists, we must keep in mind that we can only observe it indirectly trough a limited number of samples in practice.

Without limiting the generality of the model, we abstract from user mobility. Hence, the user population (the set of potential contributors, e.g. the members of a social network) can be modeled as a set of coordinates $U = \{(x_1, y_1), (x_2, y_2), (x_3, y_3), \ldots, (x_z, y_z)\}$, reflecting the locations of the users in the time span of the analysis. The set of meme occurrences $C \subseteq U$ can be considered as outcome of a random variable gained from evaluating $f(x, y)$ for all users U.

An important consideration is that in most analyzed scenarios, it is very unlikely that we have access to the entire set C of meme occurrences and the entire set U of user locations. For instance, the locations of all Twitter users are not available and not all occurrences of a hashtag are geotagged. Hence, we should assume that only a subset $C_n \subset C$ is available for analysis that reflects the communication behavior of only $n \leq z$ users. We denote the true Entropy, Focus and Spread based on the entire (but not observable) set of occurrences C as E^C, F^C and S^C respectively. The true frequency count in the zone $i \in R$ will be denoted as P_i^C. We can now formulate two basic requirements:

- **Requirement P1:** The true but not observable measures E^C, F^C and S^C should reflect characteristics of the geographic relevance function $f(x, y)$ and should not be biased too strongly by the distribution of the users U.
- **Requirement P2:** The observable measures E, F and S should not deviate substantially from E^C, F^C and S^C.

It is obvious that both requirements are not fulfilled entirely by the measurements under examination in a sense that they are unbiased by the distribution of the population and the selection of samples. However, the strength of these biases has not been analyzed before and therefore it is also not known to which degree the considered measurements are appropriate for analyzing the geographic relevance of geotagged social media contents at all.

3.2   Sampling Dataset

Geographic data of a large set of randomly collected tweets are considered as a reasonable estimate of the varying density of the user population across different regions. We collected a corpus of over 4 million geo-tagged tweets between 21th of June 2013 and 5th of August 2013. We only considered a single tweet per user (the first one in our data set) since our aim is to model the user density and not the tweet density. This resulted in a set of 2,620,058 coordinates, which is denoted as real data set in the remainder of this paper. Additionally, we generated a reference data set of about the same number of random coordinates which are uniformly distributed on the earth, which we refer to as baseline data set.

3.3   Simulation of Spatial Distributions

To assess the measures with respect to the requirements, we generate sample sets of meme occurrences based on the circular model using Monte Carlo simulations. To apply the circular model with radius r, first a random center point Cp is chosen. It is assumed that $f(x, y) = k$, with k being a constant, for all (x, y) with $D((x, y), Cp) < r$ and $f(x, y) = 0$ otherwise. Instead of defining k, we specify the number of samples n and generate meme occurrences in the following way: (1) A random point from the sampling data set is drawn and considered as center point Cp; (2) n points with $D((x, y), Cp) < r$ are randomly drawn from the sampling data set (Fig. 2).

In a next step (Sect. 4), we then evaluated how well Focus, Entropy and Spread reflect the concentration of the geographic relevance represented by the radius r. Independent from the location of Cp, larger radii should be indicated by a higher Entropy, a lower Focus and a higher Spread. These relationships should not be lost due to the influence of the user population and the sample size since, otherwise, the measures would not be able to reflect the geographic relevance even in the most basic scenario (circular model).

Keep in mind, that due to effects of real geography, it could be possible that the drawn Cp is located in an isolated user population cluster, i.e. on an island. In this case, the samples will be always strongly concentrated in a single location, even though the radius r is increased. In such a case, increasing radii cannot be captured by spatial measurements at all and it would not be an appropriate scenario to assess the quality of spatial measurements. To avoid such situations, criteria were defined to remove these outliers.

3.4   Region Partition

In order to measure Focus and Entropy, we apply Leopardi's algorithm to divide the globe surface into a certain number of tiles with equal area as unit regions [6].

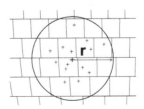

Fig. 2. The distribution area of a meme is modeled as circular area with radius r. The figure also shows the zonal partitioning according to Leopardi's algorithm (cf. Sect. 3.4)

However, the size and boundaries of unit regions will affect the frequency counts and hence the results of the measurements, also known as modifiable areal unit problem (MAUP) in spatial analysis [8]. So it is worth noting the choice of a proper scale of unit. When studying real social media contents, this choice usually depends on the intended level of analysis and interpretation, i.e., whether one is studying the distribution across countries, cities, or postcode districts, etc. Since the main interest of our tests is not concrete values but the variation and stability of measuring results, the analysis is less sensitive to the scale of unit regions. Yet, we still need to choose a scale so that unit regions are neither too big (bigger than the area of distribution) nor too small (too few points in each region). In this study, we set the number of unit regions to 10000, with an area of about $50000\,\text{km}^2$ for each region.

It should be noted that the position of a center point Cp in relation to the zone boundaries may influence spatial measurements and hence, observed variances among generated sets of meme occurrences with a fixed radius r might be caused by the underlying zone instead of the user population density. We use the baseline data set for being able to isolate zone effects in our simulations.

4   Analysis of Existing Measures

We will discuss and analyze previously stated requirement P1 in Sect. 4.1 and requirement P2 in Sect. 4.2

4.1   P1 - Dependency on User Distribution

Simulation Setting. In the basic circular model, the shape of $f(x, y)$ is specified by the center point Cp and the radius r. If the measures are not dependent on the underlying distribution of the user population, the measured values should be influenced only by r and not by the location of Cp within this setting. Interpreted as measure for geographic relevance, Focus should drop with increasing r, since it measures the tendency of concentration in a single area. Entropy should increase with increasing r, since it measures the tendency of dispersion in multiple areas. Spread should increase as well, since it measures the spatial coverage of the meme. If these correlations are low, the measures are not appropriate to draw conclusions about the geographic relevance, even in the most basic scenario (the circular model).

Hence, we generated a series of sets of meme occurrences as described in Sect. 3.2 starting with a radius of r=150 km and increasing the radius in 50 Km steps up to $r = 1150$. For each of the 20 different radius values, we conducted 30 Monte Carlo runs, each time selecting a different random center point Cp. With 500 samples per iteration we chose a number that is large enough to avoid sample size effects (P2) but still allowed us to conduct the series of simulations in a reasonable time.

Beside the distribution of the population, the MAUP could also have strong impact on the comparability of two sets of meme occurrences with different center points. To isolate this effect, we generated a second series in which we draw the user locations from the uniformly distributed baseline data set.

Results. The plots shown in Fig. 3 summarize the results and can be interpreted as follows: Each dot represents an individual set of meme occurrences consisting of 500 samples. The x-axes represent the measured values of Focus (first row), Entropy (second row) and Spread (third row). The y-axes are given by the corresponding radii r of the set of meme occurrences from which the values result. The plots in the first column result from the uniform baseline user distribution. In this idealized scenario, each measured value allows us to draw relatively precise conclusions about the radius r, because at each location on the x-axes, the values on the y-axes are spread only within small intervals. The plots in the second column show that the population density has a strong effect on the measures which masks the influence of r. For example, the results indicate that if we measure a Focus of 0.4, it could result from memes with strong local concentration (with a radius of 200 km) as well as from memes that are spread within a much wider area (with a radius above 1000 km) if we take a realistic setting into account. The experiments allow to draw the following conclusions:

- In the baseline scenario, the measured values correlate well with the radii r, with not much differences between the zone-based measures and Spread. Hence, effects caused by MAUP (i.e., whether the center point Cp is close to a zone center or a zone border) seem to be insignificant in the considered parameter range.
- In the realistic scenario, all three measurements are strongly influenced by the distribution of the user population. If we apply the measurements to compare two sets of meme occurrences, differences in the values are only indicative if they are considerably large. While, e.g., a Focus value of 0.9 indicates indeed a higher concentration (i.e., a smaller r) than a Focus value of 0.1, the measures seem not to be suitable for, e.g., a precise ranking.
- The experiment indicates that Spread seems to be most robust against effects of the user population density, while Focus seems to be most prone to biases.
- We also analyzed the mean values of the sets of meme occurrences for each radius r and the confidence intervals. Interestingly, we found no indications for systematic errors, i.e., that larger radii caused the average Focus to increase or the average Spread or Entropy to drop (cf. example in Sect. 1). Consequently, studies based on averaging over a large number of memes should not be severely biased by the distribution of user population.

4.2   Influence of Sample Size

Simulation Setting. With respect to the influence of the sample size, we want the measure to not become unstable and unreliable for a small amount of samples. In other words, smaller numbers of samples n should not cause significant deviations of the measures if the other parameters are fixed. Hence,

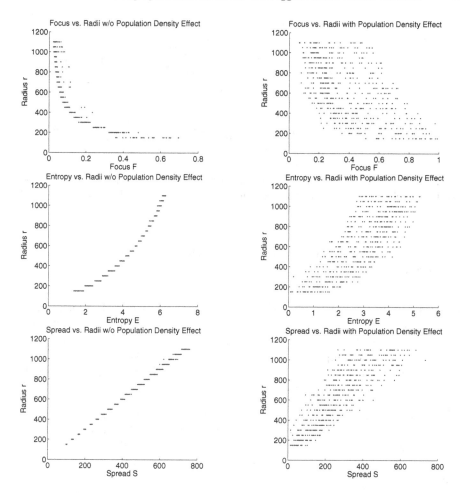

Fig. 3. Under the assumption of a uniformly distributed user population, the measured values and the radii of the distribution area correlate well (left column). Consideration of a realistic user distribution adds strong noise (right column) and the measured values do not reflect geographic relevance accurately.

we generate multiple set of meme occurrences with fixed r and varying n to analyze the stability of the measurements.

The examination of the effect of different sample sizes is especially meaningful for measures such as Focus and Entropy, because according to their definitions, the minimum value of Focus ($1/n$) and the maximum value of Entropy ($log_2 n$) are both dependent on the total number of data points.

To exclude the influence of population density, we conducted sample size tests based on the uniform baseline data set. We chose three different radii: 150 km, 1000 km and 2000 km. For each radius, we generated sets of meme occurrences ranging from 5 samples to 295 samples in steps of 10. In each step,

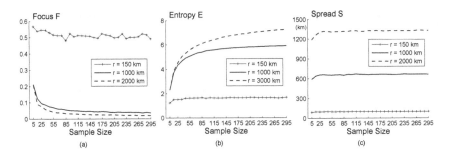

Fig. 4. (a) Focus, (b) Entropy and (c) Spread for sets of meme occurrences created with the baseline data set with various sample sizes, radii r=150 km, r=1000 km and r=2000 km

we generated 30 sets of memes and computed the average Focus, Entropy and Spread respectively.

Results. In Fig. 4 the results of our evaluation are depicted, which reveal strong systematic errors for Focus and Entropy. Mean Focus increases and mean Entropy decreases with the drop of sample size. These trends are particularly significant when the sample size is small (under 50 in our tests). Values of Spread are relatively stable, but a trend of increase can still be observed when the sample size is smaller than 50. The experiment indicates:

- If applied to small numbers of samples, there are strong tendencies of the measured Focus to overestimate the true Focus F^C. The results indicate that at least 50 or better 100 samples are required. Still, large differences in the number of samples can limit the comparability of two sets of meme occurrences.
- Even if more than 100 samples are used, the measures are biased by the sample size, which is especially a problem when Entropy is used and the number of samples in the individual set of meme occurrences differ strongly (e.g., 100 samples vs. 1000 samples). In such cases, an upper bound for the number of samples should be defined and if this number is exceeded only a subset of the samples should be used.
- If more than approximately 25 samples are used, Spread seems not to be significantly biased by the number of samples. However, the variances (not shown in the plots) of the individual sets of meme occurrences are much higher if 25 samples are used instead of 100 or more.

5 Reducing Biases

In Sect. 5.1, we first propose adjustments for Focus and Entropy to make them more robust against non equal user distributions (requirement P1) and we explain why there is no convincing adjustment to mitigate the effect of varying sample sizes (requirement P2). Second, we propose an optimized method for calculating the Spread in Sect. 5.2 which is less affected by outliers, which can otherwise completely determine the implications of the measure. By means of an assessment, based on real Twitter data collected over a time span of six weeks, we demonstrate under which situations the statements made by the original and adjusted Spread measure deviate the most.

5.1   Focus and Entropy - Inherent Limitations

Requirement 1. The number of occurrences $|O_i|$ in zone i depends on both, the number of users N_i in the zone i and the characteristic of $f(x, y)$ in zone i. The aim is to eliminate or at least to decrease the influence of N_i for a more accurate measurement and better comparability of the geographic relevance. We first consider the ideal case in which the same number of users N is located in each zone. For now, we do not take the sample size problem into account and assume that all user locations and meme occurrences are known. Since the zone-based measures only capture how the samples are distributed among the zones and not the distribution within the zones, we furthermore assume that the probability that a user contributes to the meme only depends on the zone i and not on the exact coordinates, i.e., $f(x, y)$ has a constant value in each zone i. We will denote this zone-based probability as f_i. The number of samples $|O_i|$ in zone i has the expected value $E(|O_i|) = f_i N$, which will be approximated well by $|O_i|$ if the number of users N in each zone i is large enough (law of large numbers). Under this assumption, the relative frequency P_i does not depend on the number of users N and reflects only the relative value of f_i which we denote as P_i^f:

$$P_i = \frac{|O_i|}{|O|} = \frac{|O_i|}{\sum_{j \in R} |O_j|} \approx \frac{f_i N}{\sum_{j \in R} f_j N} = \frac{f_i}{\sum_{j \in R} f_j} = P_i^f \tag{5}$$

Hence, under the assumption of a uniformly distributed user population and an appropriately large number of users in each zone, P_i^f can be considered as the ground truth, undistorted by the population density, that is approximated by P_i. Thus, to approximate P_i^f if the number of users N_i in each zone i is *not* constant, we can use P_i' as follows instead of the relative frequency P_i:

$$P_i' = \frac{\frac{|O_i|}{N_i}}{\sum_{j \in R} \left(\frac{|O_j|}{N_j} \right)} \approx \frac{\frac{f_i N_i}{N_i}}{\sum_{j \in R} \left(\frac{f_j N_j}{N_j} \right)} = P_i^f \tag{6}$$

P_i' approximates the relative value of f_i and abstracts from the distribution of the users, by dividing f_i simply by the number of users in this zone. For versions of Focus and Entropy that are independent from the user distribution, we can use P_i' instead of P_i, if the number of samples in each zone is large enough. However, if the number of samples is too small, P_i' will not approximate P_i^f well and, moreover, individual samples can have a much higher influence on

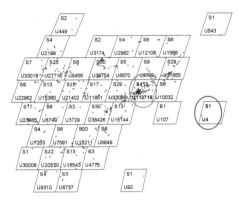

Fig. 5. Visualization of a real world twitter data set where every dot represents one meme. The total number of meme occurrences is depicted at the top edge, the total number of potential users at the bottom edge. The original Focus determines the focus area to be the gray encircled area, while an intuitive normalization method, by dividing by the number of users per region, is prone to outliers and yields the blue encircled area in this scenario (Color figure online).

the results in comparison with the original measures if they appear in zones with few users. Hence, we risk boosting random effects.

A real world scenario, highlighting this shortcoming is visualized in Fig. 5. The focus of the original measure is determined by the gray encircled area, due to having the most meme occurrences. Normalizing the Focus measure solely by the number of users, however, yields a completely different picture, namely a new focus area, the one encircled in blue. This is reasoned by the fact that there are only four potential users (U4) within the blue encircled area and one published meme (S1) compared to 113718 potential users with 113 memes in the gray encircled area. In our last publication [11] we consequently presented an approach, which on the one hand normalizes on the basis of the underlying user distribution, but on the other hand tries to avoid creating such outliers by not solely considering the actual number of potential users in a specific area, but also the average number of potential users. This way we were able to achieve a tradeoff between the risk of boosting random effects and the goal of abstracting from the user distribution. Further limitations, for instance based on too small sample sets, however, remain for both measures, the Focus and Entropy and can not be corrected as is discussed in the following.

Requirement P2. According to the definition of the measures, minimum Focus and maximum Entropy depend on the number of samples. When measuring a set C_n with n occurrences, the Focus values F will be in the range $[\frac{1}{n}, 1]$ and the Entropy values E in the range $[0, log_2 n]$. Thus, the possible values of both measures depend on the sample size. Obviously, we could simply normalize both measures to an interval of $[0, 1]$ by using the normalized Shannon Entropy $E_{norm} = E/(log_2 n)$ [5] and a normalized Focus measure $F_{norm} = (F - \frac{1}{n})/(1 - \frac{1}{n})$.

However, this does not solve the problem. We would just induce another bias since even if the minimal measurable Focus is $\frac{1}{n}$ and the maximal measurable Entropy is $log_2 n$, the true Focus F^C and Entropy E^C might still lie above or below respectively. Therefore, it is an inherent limitation of these measures that they cannot capture the entire range of possible values when the subset is smaller than the underlying entire set, hence resulting in systematic errors. We can only ensure that an appropriate sample size is used to avoid too strong biases.

5.2   Spread

Zone-based measures have the advantage that the number of users in each zone can be counted and, hence, the user density can be easily incorporated into the model. In order to abstract from the user distribution, we can formulate a zone-based approximation S' of the Spread measure S by mapping all samples within a zone i to the mean point of the zone m_i. Then, S' can be calculated based on the zone frequencies as follows:

$$S' = \frac{\sum_{i \in R} |O_i| D(m_i, Mp)}{|O|} = \sum_{i \in R} \left(\frac{|O_i|}{|O|} D(m_i, Mp) \right) = \sum_{i \in R} P_i D(m_i, Mp) \quad (7)$$

Having formulated a zone-based approximation of Spread, we can now apply the same strategy as described above for Entropy and Focus to abstract from the user distribution. Hence, we can use relative zone frequencies P_i' instead of the absolute zone frequencies P_i. Then, we also need to consider the P_i' as weights to calculate the weighted mean point, denoted as Mp'. Instead of mapping all samples in a zone to the mean point of the zone, we can use the average distance of the samples in a zone to the weighted mean point Mp' to improve the accuracy. This leads to the following adjusted Spread measure:

$$S' = \sum_{i \in R} \left(P_i' \sum_{o \in O_i} \frac{D(o, Mp')}{|O_i|} \right) \quad (8)$$

Quite obviously, this adjusted measure, however, still struggles with the same limitation than previous measures, namely that individual samples can have a determining impact on the overall evaluation if they occur in zones with a low user population (c.f. Fig. 5). Based on this insight our subsequent optimization follows the general idea of enlarging or aggregating zones until a minimum amount of samples in the newly generated zones is achieved. Even though, we only apply this concept to Spread in the following, a similar, but more complex approach could be followed for Focus and Entropy as well.

In particular, we aggregated zones with a similar distance to the mean point until we reached a minimum of five samples per zone. Calculating the distance between the center of zones and the mean point is quite easy, the threshold of five samples is chosen arbitrarily. The general approach is visualized in Fig. 6. The zones a, b and c contain too few samples and are thus aggregated into one combined zone whose sample size exceeds the defined threshold. The order

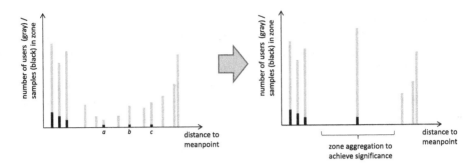

Fig. 6. General aggregating principle of the adjusted Spread measure: Originally, the single sample in zone a, in combination with the low user population, would completely dominate the Spread measure and potentially distort implications. Aggregating these zones (the number of samples as well as the population) has the potential to reduce the effect of these outliers as is visualized.

of aggregating zones can hereby solely be determined by the distance to the mean point, without having to consider whether, for instance, two zones are neighboring, which is due to the characteristic of the Spread measure to represent the mean distance of all meme occurrences to their geographic mean point. Hence, starting from the nearest to the farthest zone, it is checked whether sufficient samples are present and in case the threshold is not exceeded the next farther zone is added until the sample threshold is reached. Adding two zones hereby results in a new zone with a new center point and the combined number of samples and users of both previous zones.

Evaluation of the Adjusted Spread Measure. First, we performed an evaluation similar to the one in Sect. 4.1. The results are hereby presented in Fig. 7. The left sub figure corresponds with the one shown in Fig. 3, representing the non-adjusted Spread measure for a realistic user distribution. The right plot corresponds with the previously presented adjusted Spread measure which results in visibly less scattering.

Twitter Data Assessment. In order to demonstrate which effect the adjusted Spread measure can have on real world assessments, we conducted an assessment based on real Twitter data and evaluated whether differences between the original and adjusted Spread measure became apparent for tweets with an obvious geographic relevance. The data was hereby collected in the time frame from 21st of June 2013 until 4th of August 2013 by first listening to the Twitter-Stream-API and identifying the hashtags which were occurring the most. In a second step we then queried the Twitter-Search-API for geotagged tweets of the identified often occurring hashtags. This way we were able to extract a total of 1816 hashtags, each with at least 100 corresponding geotagged tweets.

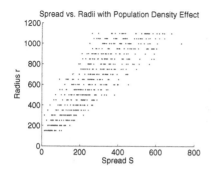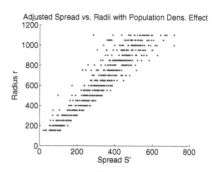

Fig. 7. Comparison between the non-adjusted (c.f. Fig. 3) and adjusted Spread measure for a realistic user distribution. It becomes apparent that the resulting radius range is clearly cut down for the adjusted Spread measure.

In order to identify hashtags with a higher geographic relevance, i.e., being of interest more locally, we matched the collected data with the geonames.org database. Our underlying assumption for doing so was that tweets with a city hashtags should be of more interest within a smaller region, e.g., the corresponding city, compared to a hashtag of, e.g., a brand or a movie. Our resulting data set, however, still contained a lot of tweets which were obviously not related to a city, state, or country. The reason for this are linguistic ambiguities, for instance that the first name of the famous basketball player *Kobe* Bryant is also a city in Japan or that *Opera* is as well a small community in Italy. In order to filter out these ambiguous hashtags and solely focus on hashtags representing cities or states, we employed several further optimizations.

1. We ignored every hashtag which corresponds with a city of less than 250000 inhabitants. The reason for this is that our study has shown that often tweeted terms, such as *gym, male,* or *top* correspond with rather small cities which are, however, not meant in this case.
2. We filtered out every city related hashtag if there is a second city with the same name and more than 5000 inhabitants. The reason for this is that we can no longer assume that the considered hashtag only has a local geographic relevance.

As a result, we cut down our considered data set to only 59 hashtags, each corresponding to a rather big city where we can now confidently assume a higher than average geographic relevance. Considering the previously mentioned shortcomings of the original Spread metric, namely its "vulnerability" towards regions with low user populations, we now expect the adjusted Spread measure to be able to better depict a higher geographic relevance of the remaining city hashtags. In order to verify this, we applied the Spread measure as well as the adjusted Spread measure on our complete data set and sorted the results in ascending order, resulting in higher geographic relevance hashtags being in front. In a subsequent step, we then calculated the average position of the 59 city hashtags with

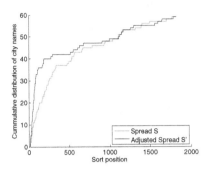 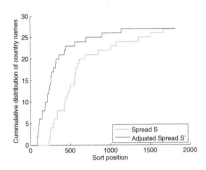

Fig. 8. Comparison between the relative positions of city and country hashtags in a sorted list based on the original and adjusted Spread measure. The adjusted Spread measure clearly depicts these city and country hashtags a higher geographic relevance.

both Spread measures and achieved an average position of 460 with the original Spread measure and a position of 369 with the adjusted Spread measure, suggesting a higher geographic relevance. Applying the same evaluation method to countries resulted in an average position of 614 for the original Spread measure and 313 for the adjusted one, again implying a higher geographic relevance compared to other considered hashtags. In both cases, the adjusted Spread measure thus assessed the city and country hashtags a relatively higher geographic relevance which corresponds with the expected result. Figure 8 confirms this insight, by highlighting the difference in positions for city and country hashtags. It becomes obvious that the adjusted Spread measure clearly assigns these city and country hashtags a higher geographic relevance.

Let us now consider two individual hashtags in order to demonstrate in which cases the original and adjusted Spread measure differ the most. Figure 9 depicts the worldwide occurrences of the hashtag *Chihuahua* which represents a city in Mexico (where the arrow is pointing to) and its corresponding state, but also a famous dog breed. With the original Spread measure this hashtag was on position 949 due to individual hashtags occurring all around the world and also in regions with low population. Due to the combination of these low population regions, the position in the geographic relevance list increases to 103. The adjusted Spread measure hence allows us to discover a more local geographic relevance (within Mexico), which would have been otherwise overlooked.

The largest difference for non-city hashtag occurred for the hashtag *Takenote* which raised from 1567 with the original Spread measure to 209 with the adjusted one. The original position of 1567 out of the total of 1816 rather indicated a global geographic relevance, compared to the new position of 209 which indicates a rather local relevance. Figure 10 visualizes again every tweet which we recorded with regard to this hashtags and indeed, there is a rather high geographic relevance visible in Malaysia which the original Spread measure did not carry weight.

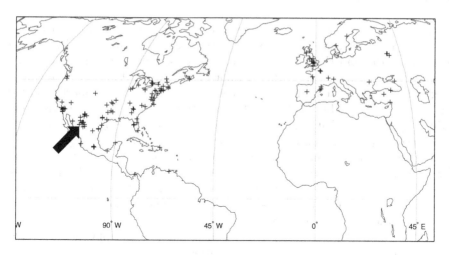

Fig. 9. Geographic occurrences of the hashtag *Chihuahua*. On the one hand it represents a city and state within Mexico (where the arrow is pointing to) and on the other hand a famous dog breed. The adjusted Spread measure carries more weight towards a higher geographic relevance, compared to the original one.

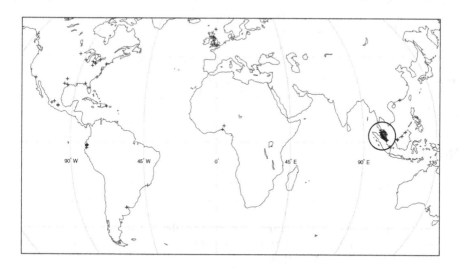

Fig. 10. Geographic occurrences of the hashtag *Takenote*. The adjusted Spread measure carries more weight towards a higher geographic relevance in Malaysia (encircled area), compared to the original Spread measure which rather indicated a more global relevance.

6   Conclusion

In this paper, we have analyzed to which degree current measures that are employed for the geospatial analysis of social media are influenced by the distribution of the underlying user population and sample size. Our results indicate that neither Spread, nor Entropy, nor Focus are able to solely characterize geographic relevance, but that the outcome is mixed up with impact from the underlying user distribution as well. Overall, our conducted Monte Carlo simulations demonstrate that too few samples and effects caused by the distribution of the user population can distort spatial measures so that they do not reflect geographic relevance accurately. Even though, we based our analysis only on three selected measures, our results indicate that the described problems are not specific to the measures under examination but apply to other spatial measures as well. Based on our experiments, we provided concrete hints that indicate in which scenarios the measures are still applicable.

We proposed adjustment methods for the Focus and Entropy measures that decrease the influence of the user distribution. Hence, the adjusted measures reflect geographic relevance more accurately, even though, the biases are not eliminated entirely. Furthermore, we showed that the sample size problem leads to systematic errors that cannot be adjusted easily because it is not possible to draw conclusions about the geographic relevance at all if the number of samples is too small. Furthermore, we proposed an aggregation scheme for the Spread measure which mitigates the influence of outliers which can otherwise completely determine the Spread measure. By means of an assessment, based on real world data, we showed the importance of this adjustment and demonstrated how statements could otherwise be distorted.

The findings presented in this paper suggest reconsidering current beliefs on the temporal evolution of social media content. We conclude that future studies should reinvestigate spatio-temporal patterns of social media distribution with much more dense data sets that provide a sufficient number of samples in each time interval.

References

1. Backstrom, L., Kleinberg, J., Kumar, R., Novak, J.: Spatial variation in search engine queries. In: Proceedings of the 17th International Conference on World Wide Web, pp. 357–366. ACM (2008)
2. Brodersen, A., Scellato, S., Wattenhofer, M.: Youtube around the world: geographic popularity of videos. In: 21st International Conference on World Wide Web, WWW 2012, pp. 241–250. ACM (2012)
3. Ding, J., Gravano, L., Shivakumar, N.: Computing geographical scopes of web resources. In: Proceedings of the 26th International Conference on Very Large Data Bases, VLDB 2000, pp. 545–556. Morgan Kaufmann Publishers Inc., San Francisco (2000)

4. Kamath, K.Y., Caverlee, J., Lee, K., Cheng, Z.: Spatio-temporal dynamics of online memes: a study of geo-tagged tweets. In: 22nd International Conference on World Wide Web, WWW 2013, pp. 667–678. International World Wide Web Conferences Steering Committee (2013)
5. Kumar, U., Kumar, V., Kapur, J.N.: Normalized measures of entropy. Int. J. Gen. Syst. **12**(1), 55–69 (1986)
6. Leopardi, P.: A partition of the unit sphere into regions of equal area and small diameter. Electron. Trans. Numer. Anal. **25**, 309–327 (2006)
7. Li, L., Goodchild, M.F., Xu, B.: Spatial, temporal, and socioeconomic patterns in the use of twitter and flickr. Cartography Geogr. Inf. Sci. **40**(2), 61–77 (2013)
8. Openshaw, S.: The Modifiable Areal Unit Problem. Concepts and Techniques in Modern Geography. Geo Books, Norwick (1984)
9. Petrovic, S., Osborne, M., McCreadie, R., Macdonald, C., Ounis, I., Shrimpton, L.: Can twitter replace newswire for breaking news? In: Seventh International AAAI Conference on Weblogs and Social Media. The AAAI Press (2013)
10. Scellato, S., Noulas, A., Lambiotte, R., Mascolo, C.: Socio-spatial properties of online location-based social networks. In: ICWSM (2011)
11. Wang, X., Gaugel, T., Keller, M.: On spatial measures for geotagged social media contents. In: Proceedings of the 5th International Workshop on Mining Ubiquitous and Social Environment (2014)

Formation and Temporal Evolution of Social Groups During Coffee Breaks

Martin Atzmueller[1]([✉]), Andreas Ernst[2], Friedrich Krebs[2], Christoph Scholz[3], and Gerd Stumme[1]

[1] Knowledge and Data Engineering Group, Research Center for Information System Design, University of Kassel, Wilhelmshöher Allee 73, 34121 Kassel, Germany
{atzmueller,stumme}@cs.uni-kassel.de
[2] Center for Environmental Systems Research, University of Kassel, Wilhelmshöher Allee 47, 34117 Kassel, Germany
{ernst,krebs}@usf.uni-kassel.de
[3] Department Energy Informatics and Information Systems, Fraunhofer Institute for Wind Energy and Energy System Technology, Königstor 59, 34119 Kassel, Germany
christoph.scholz@iwes.fraunhofer.de

Abstract. Group formation and evolution are prominent topics in social contexts. This paper focuses on the analysis of group evolution events in networks of face-to-face proximity. We first analyze statistical properties of group evolution, e.g., individual activity and typical group sizes. After that, we define a set of specific group evolution events. These are analyzed in the context of an academic conference, where we provide different patterns according to phases of the conference. Specifically, we investigate group formation and evolution using real-world data collected at the LWA 2010 conference utilizing the Conferator system, and discuss patterns according to different phases of the conference.

1 Introduction

An important goal of social sciences is to reach a theoretical understanding of the process of group formation and evolution of humans [24]. Typically, in such contexts the analysis is enabled using empirical studies of human behavior. However, until recently, such studies were very costly and time-consuming, especially for larger groups: Here, the individual behaviors of a larger group of people had to be observed – for a longer time period in a not too small area. Indeed, now – with the rise of social networking sites such as Second Life or Facebook – the situation in data collection has changed significantly. With such systems in place, the situation is quite different, as it has become much easier to track the individual behavior of users. While there have been results indicating that online connections relate to offline connections in specific contexts, e. g., [30], it has also been argued that the behavior within these online platforms differs in many cases from the offline behavior and its inherent structures. Strong ties, for example, seem to correlate better than weak ties [30], but e. g., also only a small share of friends in Facebook are really close connections, i. e., friends in the offline world [32,53].

© Springer International Publishing Switzerland 2016
M. Atzmueller et al. (Eds.): MSM, MUSE, SenseML 2014, LNAI 9546, pp. 90–108, 2016.
DOI: 10.1007/978-3-319-29009-6_5

With the further development of sensor technology, however, it has become possible to track the behavior of individuals also in the offline world. Using suitable sensors, collecting data from large(r) groups has become possible. In our work, we will make use of RFID technology to track not only the location of individuals, but also to observe their communication behavior [16].

We utilize data of the Conferator[1] system [5] – a social conference guidance system for enhancing social interactions at conferences. Conferatorapplies active RFID proximity tags developed by the Sociopatterns collaboration.[2] In particular, these tags allow the collection of human face-to-face proximity. For our analysis, we utilize data that has been collected at the academic conference LWA 2010.[3] For the event, the participants of the conference were wearing the RFID tags for three days, at all times during the conference time.

Based on these data, we have performed an analysis of the formation and breakup of groups. Our contribution can be summarized as follows:

1. We provide a formal model of group evolution in networks of face-to-face proximity and present a definition of different group evolution events.
2. We then consider social behavior of individuals and specifically analyze the evolution of social groups:
   (a) We provide a statistical analysis of individual activity and typical group sizes during conference phases.
   (b) Second, we investigate the temporal evolution of the proposed group evolution events throughout the conference and especially during the coffee breaks. As a result, we observe and discuss typical communication and activity patterns during these social events.
3. We analyze these patterns and characteristics and discuss quite clear-cut differences between conference sessions, coffee breaks, poster sessions, and free time.

The rest of this paper is structured as follows: Sect. 2 discusses related work. Section 3 describes the RFID hardware setting and gives a detailed overview on the collected real-world datasets. Section 4 describes the formalization of social groups and group transitions. After that, Sect. 5 presents the analysis. Finally, Sect. 6 summarizes our results and discusses future work.

2 Related Work

In this section, we discuss related work concerning the analysis of human contact behavior and the analysis of groups. We start with a detailed overview on the analysis of human contact behavior.

---

[1] http://www.conferator.org.
[2] http://www.sociopatterns.org.
[3] http://www.kde.cs.uni-kassel.de/conf/lwa10.

2.1   Human Contact Behavior

The analysis of human contact patterns and their underlying structure is an interesting and challenging task in social network analysis. Eagle and Pentland [26], for example, presented an analysis using proximity information collected by bluetooth devices as a proxy for human proximity. However, given the interaction range of bluetooth devices, the detected proximity does not necessarily correspond to face-to-face contacts [16], as also confirmed by Atzmueller and Hilgenberg [8]. The SocioPatterns collaboration developed an infrastructure that detects close-range and face-to-face proximity (1–1.5 m) of individuals wearing proximity tags with a temporal resolution of 20 s [23]. This infrastructure was also deployed in other environments in order to study human contacts, such as healthcare environments [28,35], schools [51], offices [21] and museums [29]. Here, first analyses concerning group contact evolution have been reported in [15], focusing on the temporal evolution of smaller groups (up to size four).

Another approach for observing human face-to-face communication is the Sociometric Badge.[4] It records more details of the interaction but requires significantly larger devices. Besides these two approaches, there is, up to our knowledge, no single empirical study in the social sciences that resulted in a history of all conversations of some event, where, for each face-to-face conversation, the names of the respective dialogue partners are stored together with exact time stamps for start and end of the conversation.

The SocioPatterns framework also provides the technical basis of our Conferator [3–5] system. In this context, Atzmueller et al. [6] analyze the interactions and dynamics of the behavior of participants at conferences; similarly, the connection between research interests, roles and academic jobs of conference attendees is analyzed in [34]. Furthermore, the predictability of links in face-to-face contact networks and additional factors also including online networks have been analyzed by Scholz et al. [45,46].

2.2   Analysis of Groups

Groups and their evolution are prominent topics in social sciences, e. g., [24,33,56]. Wasserman and Faust [57] discuss social network analysis in depth, and provide an overview on the analysis of cohesive subgroups in graphs, both outlining methods for structural analysis, e. g., [27,41] as well as for obtaining compositional descriptions, e. g., [1,2]. Social group evolution has been investigated in a community-based analysis [42] using bibliographic and call-detail records. Backstrom et al. [14] analyze group formation and evolution in large online social networks, focussing on membership, growth, and change of a group. Furthermore, Brodka et al. [19,20,44] investigate group formation and group evolution discovery and prediction in social networks.

In contrast to the approaches above, this paper focuses on networks of face-to-face proximity at academic conferences: We extend the definitions for group

---

[4] http://hd.media.mit.edu/badges.

formation and evolution in a fine-grained analysis and investigate the impact of different phases at a conference. Furthermore, we do not necessarily focus on groups defined by a dense graph-structure, but analyze respective groups that are connected by face-to-face contacts. To the best of the authors' knowledge, this is the first time that such an analysis has been performed using real-world networks of face-to-face proximity.

3   Face-To-Face Contact Data

In this section, we summarize the framework used for collecting face-to-face contact networks, before we briefly describe the Conferator system.

3.1   RFID Setup

At LWA 2010 we asked participants to wear the active SocioPatterns RFID devices (see above), which can sense and log the close-range face-to-face proximity of individuals wearing them. Using the UBICON framework [3,4] for data collection, this allows us to map out time-resolved networks of face-to-face contacts among the conference attendees. In the following, we refer to the applied active RFID tags as *proximity tags*.

A proximity tag sends out two types of radio packets: Proximity-sensing signals and tracking signals. Proximity radio packets are emitted at very low power and their exchange between two devices is used as a proxy for the close-range proximity of the individuals wearing them. Packet exchange is only possible when the devices are in close enough contact to each other (1–1.5 m). The human body acts as an RF shield at the carrier frequency used for communication [23].

For estimating a face-to-face contact, we apply a similar threshold-based approach as in [23]: We record a face-to-face contact when the length of a contact is at least 20 s. A contact ends when the proximity tags do not detect each other for more than 60 s. With respect to the accuracy of the applied RFID tags, we refer to the results of Cattuto et al. [23] who confirm (1) that if the tags are worn on the chest, then very few false positive contacts are observed, (2) face-to-face proximity can be observed with a probability of over 99 % using the interval of 20 s for a minimal contact duration. This is in the range of human inter-annotator-agreement [22]. Compared to their experiments, our setup is even more conservative since we use a threshold of 60 s when determining the end of a contact. Furthermore, it is important to note that we focus on face-to-face proximity as a proxy for actual communication; due to the applied thresholds (see above), face-to-face proximity situations which include episodes that are, e. g., briefly side-by-side or over the shoulder, can typically also be captured.

The proximity tags also send out tracking signals at different power levels, that are received by antennas of RFID readers installed at fixed positions in the conference environment. These tracking signals are used to relay proximity information to a central server and also to provide approximate (room-level) positioning of conference participants, cf. [47,48]. This allows us to monitor encounters, e.g., the number of times a pair of participants is assigned to the same set of nearest readers. All the packets emitted by a proximity tag contain a unique numeric identifier of the tag, as well the identifiers of the detected nearby devices. For more information about the proximity sensing technology, we refer the reader to the website of SocioPatterns.[5] For more details on the context of the LWA 2010 conference, we refer to [6] for an in-depth presentation.

3.2 Conferator Platform

The proximity tags described above provide the physical infrastructure for our social conference management system Conferator. Conferator [3–5] is a social and ubiquitous conference guidance system, aiming at supporting conference participants during conference planning, attendance and their post-conference activities. Conferator features the ability to manage social and face-to-face contacts during the conference and to support social networking. Among other features, it provides an overview on the current social events and interactions, a map for locating conference participants, a personalized schedule, and adaptive recommendation mechanisms for interesting contacts and talks at the conference.

Conferator has successfully been deployed at several events, e.g., the LWA 2010,[6] LWA 2011[7] and LWA 2012[8] conferences, the Hypertext 2011[9] conference, the INFORMATIK 2013[10] conference, the UIS 2015 workshop[11] and a technology day of the Venus[12] project. In this paper, we focus on the data obtained at LWA 2010 in Kassel.

4   Formal Model

Before we analyze the evolution of groups in face-to-face contact networks, it is necessary to give a definition of a temporal social network and a social group.

4.1   Modeling Social Groups

Let $F = ([t_1, t_2), [t_2, t_3) \ldots, [t_m, t_{m+1}))$ be a list of consecutive time windows. In this paper, all windows will have a duration of one minute. Similar to [20] we define a *temporal social network* TSN as a list of single social networks $(SN_1, \cdots, SN_m)$.

$$SN_i = (V_i, E_i), i = 1, 2, \ldots, m,$$

---

[5] http://www.sociopatterns.org.
[6] http://www.kde.cs.uni-kassel.de/conf/lwa10/.
[7] http://lwa2011.cs.uni-magdeburg.de/.
[8] http://lwa2012.cs.tu-dortmund.de/.
[9] http://www.ht2011.org/.
[10] http://informatik2013.de/.
[11] http://enviroinfo.eu/ak-uis/uis-2015.
[12] http://www.iteg.uni-kassel.de/.

where V_i is the set of all participants who had at least one face-to-face contact with some other participant within the time window $[t_i, t_{i+1})$. Two participants $u, v \in V_i$ are connected by an edge $e := (u, v)$ in E_i if they had at least one face-to-face contact within the time window $[t_i, t_{i+1})$.

We define a *social group* G in the social network $SN = (V, E)$ as a subset of vertices $G \subseteq V$ where G is a connected component of SN with $|G| > 1$. We denote the set of all social groups of SN by $\mathcal{G}$, and the set of all social groups of SN_i by $\mathcal{G}_i$.

4.2 Modeling Group Transitions

As in [20] we differentiate between the group transitions *form, merge, grow, continue, shrink, split,* and *dissolve* between two consecutive time windows $[t_i, t_{i+1})$ and $[t_{i+1}, t_{i+2})$. However, below we provide more formal and stricter definitions that allow us to classify evolution events without exceptions.

– We say that a group G *forms* in SN_{i+1}, iff $G \in \mathcal{G}_{i+1}$ and $\forall g_i \in G : \not\exists G^{'} \in \mathcal{G}_i : g_i \in G^{'}$.
– We say that groups $G_1, \ldots, G_m$ in SN_i *merge*, iff $m \geq 2$ and $\exists G \in \mathcal{G}_{i+1}$ such that $\bigcup_{i=1}^{m} G_i \subseteq G$.

We say that group G in $\mathcal{G}_i$

– *grows*, iff $\exists! G^{'} \in \mathcal{G}_{i+1} : G \subset G^{'}$,
– *continues*, iff $\exists G^{'} \in \mathcal{G}_{i+1} : G^{'} = G$,
– *shrinks*, iff $\exists! G^{'} \in \mathcal{G}_{i+1} : G \supset G^{'}$,
– *splits*, iff $\exists G_1, \ldots, G_m \in \mathcal{G}_{i+1}$, with $m \geq 2$ such that $\bigcup_{i=1}^{m} G_i \subseteq G$,
– *dissolves*, iff $\forall g_i \in G : \not\exists G^{'} \in \mathcal{G}_{i+1} : g_i \in G^{'}$.

5 Analysis

In this section, we first describe the applied dataset. After that, we provide statistical analysis results on individual activity, before we investigate group formation and evolution in detail and provide illustrating examples.

5.1 Dataset

Each link in the applied LWA 2010 network indicates physical face-to-face proximity and can be weighted by the cumulated duration of all face-to-face proximity contacts between the linked persons.

Table 1 provides a detailed overview on the dataset. As already observed in many other contexts [23,29,34] the distributions of all aggregated face-to-face contacts lengths between conference participants are heavy-tailed. More than the half of all cumulated face-to-face contacts are less than 200 s and the average contact duration is less than one minute, but very long contacts are also observed. Overall, the diameter, the average degree and the average path length of G are similar to the results presented in [6,29].

Table 2 shows statistics on the individual group evolution events for different minimum group sizes for the LWA 2010 dataset.

Table 1. General statistics for LWA 2010 dataset. In addition to the number of nodes and edges, here d is the diameter, AACD the average aggregated contact-duration (in seconds) and APL the average path length.

$|V|$	$|E|$	Avg. Deg.	APL	d (G)	$AACD$
77	1004	26.07	1.7	3	797

Table 2. Statistics on the individual group evolution events for different minimum group sizes.

	≥ 2	≥ 3	≥ 4	≥ 8
Forming	941	98	16	1
Dissolving	936	96	16	1
Merging	140	140	140	50
Splitting	146	146	146	53
Growing	839	839	461	94
Shrinking	835	835	463	83
Continuing	3951	1103	406	33

5.2 Social Behavior of Individuals

This analysis draws on assessing the quantity and the quality of contacts during the course of a conference and the respective heterogeneity of individual conference participants. We consider three different temporal phases during the conference, i. e., coffee (and lunch) breaks, conference sessions, poster session, and free time (i. e., the remaining time besides breaks and sessions).

The contact quantity provides an indicator of the networking activity of an individual while attending the conference. In a given phase of the conference, we measure contact activity by relating the number of minutes a participant attended to the number of minutes during which a contact with another participant was observed. The resulting indicator is quantified in terms of the mean number of contacts per hour of an individual participant during the respective conference phases. Figure 1 illustrates the results.

On average, individuals have 23 ($sd = 12.11$) contacts per hour during coffee breaks, 15 ($sd = 9.13$) during sessions, and 27 ($sd = 15.7$) during their free time. Differences in contacts per hour between conference phases are significant (repeated measures ANOVA with participants as within-factor and mean contacts per hour in different phases as dependent variables, $F(1.531, 105.634) = 32.216, p < .01$, Greenhouse-Geisser adjusted). Pairwise comparisons between phases using paired t tests show significant differences between session and coffee breaks ($T(74) = -6.64, p < .01$) and session and free time ($T(69) = -7.503, p < .01$). Differences between coffee breaks and free time were not significant ($T(69) = -2.009, p = .048$, adjusted alpha level $= 0.017$ (Bonferroni)).

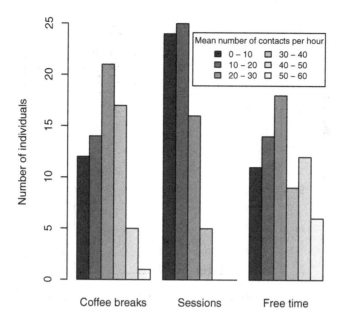

Fig. 1. Histograms of contact activities during conference phases. Except for the lowest category the cells denote right-closed and left open intervals.

These overall and pairwise results were confirmed by the equivalent nonparametric test (Friedman test, $X^2(2) = 51.686, p < 0.01$).

Unsurprisingly, during coffee breaks or free times contact activity increases compared to session times. In both phases, a majority of the participants has more than 20 and up to 60 contacts per hour. In contrast, during session time the observed number of contacts decreases to 20 or less per hour for a big majority of the participants.

5.3   Evolution of Social Groups

In the following, we first investigate group statistics, focusing on group sizes during different conference phases. After that, we investigate group evolution events in detail.

Group Statistics. While the previous analysis focused merely on the quantity of contacts by an individual, the following investigation looks at a different property of the respective conversations. Thus, we determine the size of the conversation group an individual finds himself in during a given minute of the conference. Such conversation groups correspond to connected social network components as defined in Sect. 4.

Our assumption is that being member of a larger conversation group enables an individual on the one hand to spread his thoughts and ideas more widely

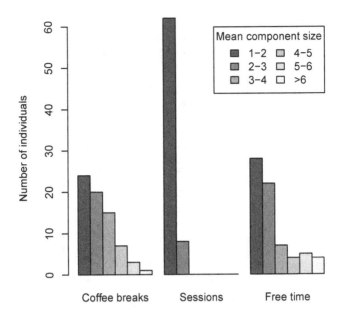

Fig. 2. Histograms of conversation group sizes during conference phases. Except for the lowest and highest categories histogram cells are right-closed and left open intervals. Note that component size 1 is included in the statistics to cover the case of solitary standing conference participants.

and on the other hand allows him to perceive more diverse contributions from other individuals. Of course it has to be noted that face-to-face conversation or very small conversation groups can likewise yield high quality information exchange. However, in the context of this paper we use the size of the component an individual participant belongs to during a given phase of the conference as a proxy for the conversation quality of the respective individual. Figure 2 shows the respective results. On average, individuals find themselves in conversation groups of size 2.72 ($sd = 1.2$) during coffee breaks, 1.55 ($sd = .36$) during sessions, and 2.74 ($sd = 1.47$) during free time. The differences between conference phases are significant (repeated measures ANOVA with participants as within-factor and mean group size in different phases as dependent variables, $F(1.61, 111.2) = 36.138, p < .01$, Greenhouse-Geisser adjusted). Pairwise comparisons between phases using paired t tests show significant differences between session time and coffee breaks ($T(74) = -8.81, p < .01$) and session time and free time ($T(69) = -7.43, p < .01$). Differences in group size between coffee breaks and free time were not significant ($T(69) = -0.88, p = .93$). These overall and pairwise results were confirmed by the nonparametric equivalent test (Friedman test: $X^2(2) = 65, p < 0.01$).

Clearly, during session times for the vast majority of individuals contacts are restricted to face-to-face (component size 2) or do not occur at all (component size 1). In sharp contrast, during coffee breaks or free times only one third of the participants remain in such small (conversation) groups while the others are found in larger groups up to size 6 and more. On the extreme end, around 10 participants are in average over all coffee breaks of the conference members of conversation groups of sizes exceeding 4. Similar circumstances are found during free time. Interestingly, despite significantly different activity patterns (see Fig. 1 above) conversations groups tend to be smaller during free times compared to coffee breaks. However, this difference is statistically not significant.

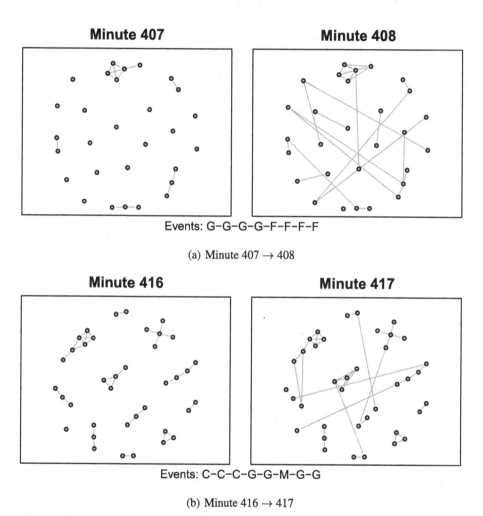

Minute 407   **Minute 408**

Events: G–G–G–G–F–F–F–F

(a) Minute 407 → 408

Minute 416   **Minute 417**

Events: C–C–C–G–G–M–G–G

(b) Minute 416 → 417

Fig. 3. Examples of group transitions at LWA 2010. The different transitions are depicted by the following annotations: C=Continuing, D=Dissolving, F=Forming, Sh=Shrinking, G=Growing, M=Merging Sp=Splitting

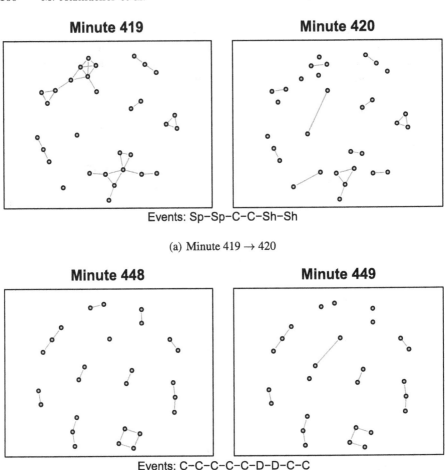

Minute 419

Minute 420

Events: Sp–Sp–C–C–Sh–Sh

(a) Minute 419 → 420

Minute 448

Minute 449

Events: C–C–C–C–C–D–D–C–C

(b) Minute 448 → 449

Fig. 4. Examples of group transitions at LWA 2010. The different transitions are depicted by the following annotations: C=Continuing, D=Dissolving, F=Forming, Sh=Shrinking, G=Growing, M=Merging Sp=Splitting

Group Transitions. In the following, we study the transition of the groups over time. Time is measured in minutes (excluding the nights). Minute 0 is 8:03 AM on Day 1 when the first signal of an RFID tag arrived, and Minute 2282 is the last signal recorded at 06:01 PM on Day 3. Day 1 ends in minute 740 with the last signal of the day on 08:23 PM; and Day 2 starts in Minute 741 at 08:14 AM with the first signal of the day. Day 2 ends in Minute 1714 with the last signal (concluding also the poster session) at 12:28 AM, and Day 3 starts in Minute 1715 at 08:34 AM. For detailed information, the conference schedule is available at http://www.kde.cs.uni-kassel.de/conf/lwa10/program.html.

We start by illustrating some typical network configurations during the first coffee break of the conference (Minutes 416–446). In doing so, we will exemplify some of the typically occurring types of transitions. At the end of a session we expect conversation groups to build up while people leave the session rooms. The figures below show the contact networks during the final minutes of the session (minutes 407 and 408) and during the official beginning of the coffee break. In the footer line of the diagrams the group evolution events identified during the transition from t to t+1 are displayed.

Between minute 407 and 408 a total of eight growing and forming events occur. People already leave the session rooms prior to the end of the session and start getting in contact. Consistently, the diagram for minute 416 illustrates that at the beginning of the break numerous groups of different sizes are established. Towards minute 417 these groups either persist, or they grow or merge respectively. Compared to the other minutes of the coffee break during these two time spans the maximum frequency of *growing* events is found. Likewise for the first time span the maximum number of forming events during the coffee break is observed.

The circumstances during the end of the coffee break and beginning of the following sessions are well illustrated by the characteristics of the transition from minute 419 to 420 and 448 to 449, see Fig. 5(a), (b), (c) and (d): The first diagram shows a case of splitting and shrinking of larger groups. The second diagram illustrates that once conversation groups have shrunk most of the remaining small groups persist and only two groups dissolve. This situation marks the maximum number of *continuing* events found during the course of the regarded coffee break. The time span from minute 448 to 449 exhibits the maximum number of *splitting* events found for the considered coffee break.

After these illustrating examples, we turn to a quantitative analysis of the group transitions:

- For our study we used different minimum group sizes. A minimum group size of $n \in \mathbb{N}$ means that we consider all groups with size greater or equal n.
- In Fig. 5, we plotted, for each transition type, the weighted sum of all its transitions between minute 0 and t.
- For each transition of one of the types *continuing, dissolving, splitting* and *shrinking*, we add $|G_t|$ to the sum. For each transition of one of the types *forming, growing* and *merging*, we add $|G_{t+1}|$ to the sum.
- At the top of the figure, we mark the different coffee breaks (shown in blue); the red bar on Day 2 indicates the poster session.

We observe that for a minimum group size of 2 the number of *continuings* is the most dominating value. The number of *continuings* decreases rapidly when we consider groups with size greater than 3 only. This means that the *continuing*-event mostly appears in groups of size 2 or 3. In addition, we note that the group transition types *forming* and *dissolving* are observed mostly for groups of size 2. To our surprise it is very unlikely that a group of size greater than 3 will *form* or *dissolve*. Considering groups of size greater than 3 the group transitions

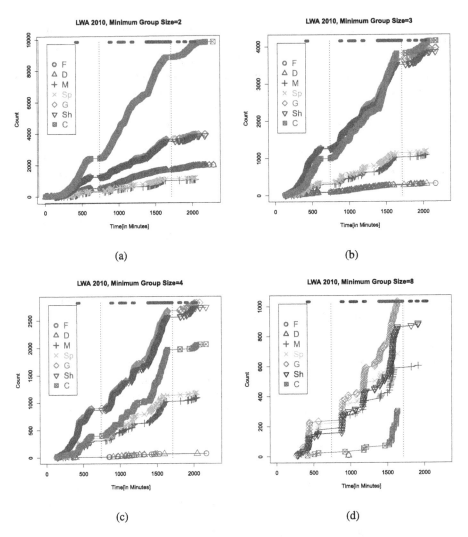

Fig. 5. Aggregated weighted occurrences of the transition types during the conference. C=Continuing, D=Dissolving, F=Forming, Sh=Shrinking, G=Growing, M=Merging Sp=Splitting. At the top of the figure, we mark the different coffee breaks (shown in blue); the red bar on Day 2 indicates the poster session (Color figure online).

growing and *shrinking* become the most dominating events. For larger groups, we observe a strong increase of *continuings* during the conference poster session.

It is interesting to see that the inverse transitions (i.e. *growing* vs. *shrinking*, *forming* vs. *dissolving* and *merging* vs. *splitting*) have almost identical curves. This is a first indicator for the hypothesis that growth and decay of communication groups are symmetric. As expected, they differ during communicative phases (coffee breaks etc.) such that the weighted sum of the increasing transition type

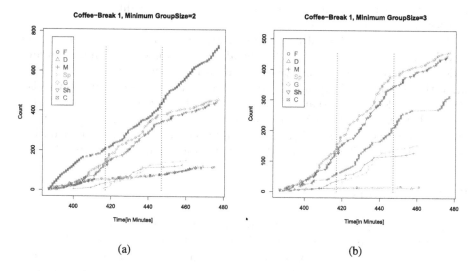

(a)

(b)

Fig. 6. Close-up of the curves in Fig. 5 around coffee break 1 for minimal groups sizes *GroupSize* = *2, 3*. For better readability, all curves start at level 0 at the left end of the diagram. The different transition types are depicted by the following annotations: C=Continuing, D=Dissolving, F=Forming, Sh=Shrinking, G=Growing, M=Merging Sp=Splitting

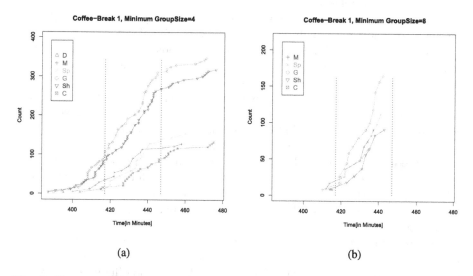

(a)

(b)

Fig. 7. Close-up of the curves in Fig. 5 around coffee break 1 for minimal groups sizes *GroupSize* = *4, 8*. For better readability, all curves start at level 0 at the left end of the diagram. The different transition types are depicted by the following annotations: C=Continuing, D=Dissolving, F=Forming, Sh=Shrinking, G=Growing, M=Merging Sp=Splitting

grows earlier during this phase, while the sum of the corresponding decreasing type grows more at the end of the phase.

For some further illustrating examples, Figs. 6 and 7 show a close-up of the global curves around the first coffee break, which started in Minute 416 and ended in Minute 446, including thirty minutes prior and after the break. Also, while the results of Figs. 6 and 7 are quite similar to those of Fig. 5, we also observe the clear trend that the most activity takes place during the coffee breaks. For example, for a minimum group size of 8 the coffee break can be detected very well (see Fig. 7(b)): Here all the group transitions take place during the coffee break. This observation does also hold for all other coffee breaks.

6 Conclusions

We have used RFID technology to investigate the structure and dynamics of real-life face-to-face social contacts. We presented a formal model of detecting group dynamics in the data providing strict definitions that allow us to classify evolution events without exceptions. As an example, we took the interactions of participants of one conference and analyzed their individual activities, as well as the characteristic and quite clear-cut differences between conference sessions, coffee breaks, poster sessions, and free time. While the data have great face validity, it will certainly be useful to validate the data provided by the RFID technology with experimental means in future research to know more about possible technical artefacts.

Furthermore, we also aim to investigate the generality of the observed phenomena by extending the analysis focusing on a set of conferences, e. g., [34,49]. Then, also subgroup and community detection methods aiming to describe such groups can provide further insights and data-driven explanations, e. g., [7,9,12]. Further fundamental issues concern the analysis of dynamics of groups and their evolution [25,31]. Also, analytical methods can then potentially be used for grounding the evaluation of such structures, e. g., [36,38]. We aim to investigate such approaches in more detail also concerning multi-layer networks, e. g., [54]. In addition, we will investigate how to embed findings on structure and dynamics into predictive methods: This includes, e. g., link prediction in such contexts [46,49,52], community analytics [7,55], and according pattern detection and modeling methods, e. g., [10,11,43]. Of course, this also extends to the semantics of user interactions [17,39,40,50], their evaluation [37] and their explanation, e. g., [13,18].

Also, at the moment, we have focussed on macro phenomena like the overall group dynamics. But the technology we use also allows for combining off-line data about individuals (like e.g. their academic role of their scientific interests) with their communication behavior at meetings. The individual history of encounters and personal acquaintances certainly plays a further role. Moreover, architectural and constructional properties of the venue can influence the formation of groups, e.g. the localization of the buffet of the conference dinner, and so forth. Further directions here also include location based group and mobility patterns.

By combining such additional knowledge with the observed real-time dynamics, we might get closer to a theory of real world face-to-face group dynamics. Such dynamics, in turn, might be taken as a proxy for the spread of information between people, or for in-depth discussions depending on the kind of groups we observe.

Acknowledgements. This work has been supported by the VENUS research cluster at the Research Center for Information System Design (ITeG) at Kassel University. We thank SocioPatterns for providing privileged access to the SocioPatterns sensing platform used for collecting the contact data.

References

1. Atzmueller, M.: Knowledge-Intensive Subgroup Mining - Techniques for Automatic and Interactive Discovery. DISKI, vol. 307. IOS Press, The Netherlands (2007)
2. Atzmueller, M.: WIREs: subgroup discovery - advanced review. Data Min. Knowl. Discov. **5**(1), 35–49 (2015)
3. Atzmueller, M., Becker, M., Doerfel, S., Kibanov, M., Hotho, A., Macek, B.-E., Mitzlaff, F., Mueller, J., Scholz, C., Stumme, G.: Ubicon: observing social and physical activities. In: Proceedings of the 4th IEEE International Conference on Cyber, Physical and Social Computing (CPSCom 2012) (2012)
4. Atzmueller, M., Becker, M., Kibanov, M., Scholz, C., Doerfel, S., Hotho, A., Macek, B.-E., Mitzlaff, F., Mueller, J., Stumme, G.: Ubicon and its applications for ubiquitous social computing. New Rev. Hypermedia Multimedia **20**(1), 53–77 (2014)
5. Atzmueller, M., Benz, D., Doerfel, S., Hotho, A., Jäaschke, R., Macek, B.E., Mitzlaff, F., Scholz, C., Stumme, G.: Enhancing social interactions at conferences. it+ti **53**(3), 101–107 (2011)
6. Atzmueller, M., Doerfel, S., Stumme, G., Mitzlaff, F., Hotho, A.: Face-to-face contacts at a conference: dynamics of communities and roles. In: Atzmueller, M., Chin, A., Helic, D., Hotho, A. (eds.) MUSE 2011 and MSM 2011. LNCS, vol. 7472, pp. 21–39. Springer, Heidelberg (2012)
7. Atzmueller, M., Doerfel, S., Mitzlaff, F.: Description-oriented community detection using exhaustive subgroup discovery. Inf. Sci. **329**, 965–984 (2016, to appear)
8. Atzmueller, M., Hilgenberg, K.: Towards capturing social interactions with SDCF: an extensible framework for mobile sensing and ubiquitous data collection. In: Proceedings of the 4th International Workshop on Modeling Social Media (MSM 2013), Hypertext 2013. ACM Press, New York (2013)
9. Atzmueller, M., Lemmerich, F.: VIKAMINE – open-source subgroup discovery, pattern mining, and analytics. In: Bie, T., Cristianini, N., Flach, P.A. (eds.) ECML PKDD 2012, Part II. LNCS, vol. 7524, pp. 842–845. Springer, Heidelberg (2012)
10. Atzmueller, M., Lemmerich, F.: Exploratory pattern mining on social media using geo-references and social tagging information. IJWS **2**(1/2), 80–112 (2013)
11. Atzmueller, M., Lemmerich, F., Krause, B., Hotho, A.: Who are the spammers? understandable local patterns for concept description. In: Proceedings of the 7th Conference on Computer Methods and Systems. Oprogramowanie Nauko-Techniczne, Krakow, Poland (2009)
12. Atzmueller, M., Puppe, F.: A case-based approach for characterization and analysis of subgroup patterns. J. Appl. Intell. **28**(3), 210–221 (2008)

13. Atzmueller, M., Roth-Berghofer, T.: The mining and analysis continuum of explaining uncovered. In: Proceedings of the 30th SGAI International Conference on Artificial Intelligence (AI-2010) (2010)
14. Backstrom, L., Huttenlocher, D., Kleinberg, J., Lan, X.: Group formation in large social networks: membership, growth, and evolution. In: Proceedings of the KDD, pp. 44–54. ACM, New York (2006)
15. Barrat, A., Cattuto, C.: Temporal Networks. Understanding Complex Systems. Springer, Heidelberg (2013). Temporal Networks of Face-to-Face Human Interactions
16. Barrat, A., Cattuto, C., Colizza, V., Pinton, J.-F., den Broeck, W.V., Vespignani, A.: High Resolution Dynamical Mapping of Social Interactions with Active RFID (2008). CoRR, abs/0811.4170
17. Bojars, U., Breslin, J.G., Peristeras, V., Tummarello, G., Decker, S.: Interlinking the social web with semantics. IEEE Intell. Syst. **23**(3), 29–40 (2008)
18. Borgatti, S.P., Mehra, A., Brass, D.J., Labianca, G.: Network analysis in the social sciences. Science **323**(5916), 892–895 (2009)
19. Kołoszczyk, B., Kazienko, P., Bródka, P.: Predicting group evolution in the social network. In: Aberer, K., Flache, A., Jager, W., Liu, L., Tang, J., Guéret, C. (eds.) SocInfo 2012. LNCS, vol. 7710, pp. 54–67. Springer, Heidelberg (2012)
20. Bródka, P., Saganowski, S., Kazienko, P.: GED: the method for group evolution discovery in social networks. SNAM **3**(1), 1–14 (2011)
21. Brown, C., Efstratiou, C., Leontiadis, I., Quercia, D., Mascolo, C.: Tracking serendipitous interactions: how individual cultures shape the office. In: Proceedings of the CSCW, pp. 1072–1081. ACM, New York (2014)
22. Cattuto, C.: Oral communication, 1 December 2014
23. Cattuto, C., Van den Broeck, W., Barrat, A., Colizza, V., Pinton, J.-F., Vespignani, A.: Dynamics of person-to-person interactions from distributed RFID sensor networks. PLoS ONE **5**(7), e11596 (2010)
24. Coleman, J.: Foundations of Social Theory. Belknap Press of Harvard University Press, Cambridge (2000)
25. Diakidis, G., Karna, D., Fasarakis-Hilliard, D., Vogiatzis, D., Paliouras, G.: Predictingthe evolution of communities in social networks. In: Proceedings of the 5th International Conference on Web Intelligence, Mining and Semantics, WIMS 2015, pp. 1:1–1:6. ACM, NewYork (2015)
26. Eagle, N., Pentland, A., Lazer, D.: From the cover: inferring friendship network structure by using mobile phone data. PNAS **106**, 15274–15278 (2009)
27. Fortunato, S., Castellano, C.: Encyclopedia of Complexity and System Science. Springer, New York (2007). Community Structure in Graphs
28. Isella, L., Romano, M., Barrat, A., Cattuto, C., Colizza, V., Van den Broeck, W., Gesualdo, F., Pandolfi, E., Ravà, L., Rizzo, C., Tozzi, A.: Close encounters in a pediatric ward: measuring face-to-face proximity and mixing patterns with wearable sensors. PLoS ONE **6**, e17144 (2011)
29. Isella, L., Stehlé, J., Barrat, A., Cattuto, C., Pinton, J.-F., Broeck, W.V.D.: What's in a crowd? analysis of face-to-face behavioral networks. J. Theor. Biol. **271**, 166–180 (2011)
30. Kibanov, M., Atzmueller, M., Illig, J., Scholz, C., Barrat, A., Cattuto, C., Stumme, G.: Is web content a good proxy for real-life interaction? a case study considering online and offline interactions of computer scientists. In: Proceedings of the ASONAM. IEEE Press, Boston (2015)

31. Kibanov, M., Atzmueller, M., Scholz, C., Stumme, G.: Temporal evolution of contacts and communities in networks of face-to-face human interactions. Sci. Chin. **57**, 1–17 (2014)

32. Kim, J., Lee, J.-E.R.: The facebook paths to happiness: effects of the number of facebook friends and self-presentation on subjective well-being. Cyberpsychol. Behav. Soc. Netw. **14**(6), 359–364 (2011)

33. Kumar, R., Novak, J., Tomkins, A.: Structure and evolution of online social networks. In: Yu, P.S., Han, J., Faloutsos, C. (eds.) Link Mining: Models Algorithms, and Applications, pp. 337–357. Springer, Heidelberg (2010)

34. Macek, B.-E., Scholz, C., Atzmueller, M., Stumme, G.: Anatomy of a conference. In: Proceedings of the ACM Hypertext, New York, pp. 245–254 (2012)

35. Machens, A., Gesualdo, F., Rizzo, C., Tozzi, A.E., Barrat, A., Cattuto, C.: An Infectious Disease Model on Empirical Networks of Human Contact: Bridging the Gap between Dynamic Network Data and Contact Matrices. BMC Infectious Diseases 13(185) (2013)

36. Hotho, A., Atzmueller, M., Stumme, G., Benz, D., Mitzlaff, F.: Community assessment using evidence networks. In: Atzmueller, M., Hotho, A., Strohmaier, M., Chin, A. (eds.) MUSE/MSM 2010. LNCS, vol. 6904, pp. 79–98. Springer, Heidelberg (2011)

37. Atzmueller, M., Benz, D., Hotho, A., Mitzlaff, F., Stumme, G.: Community assessment using evidence networks. In: Atzmueller, M., Hotho, A., Strohmaier, M., Chin, A. (eds.) MUSE/MSM 2010. LNCS, vol. 6904, pp. 79–98. Springer, Heidelberg (2011)

38. Mitzlaff, F., Atzmueller, M., Benz, D., Hotho, A., Stumme, G.: User-Relatedness and Community Structure in Social Interaction Networks (2013). CoRR, abs/1309.3888

39. Mitzlaff, F., Atzmueller, M., Hotho, A., Stumme, G.: The social distributional hypothesis. J. Soc. Netw. Anal. Min. 4(216) (2014)

40. Atzmueller, M., Stumme, G., Hotho, A., Mitzlaff, F.: Semantics of user interaction in social media. In: Ghoshal, G., Poncela-Casasnovas, J., Tolksdorf, R. (eds.) Complex Networks IV. SCI, vol. 476, pp. 13–25. Springer, Heidelberg (2013)

41. Newman, M.E.J.: Detecting community structure in networks. Eur. Phys. J. **38**, 321–330 (2004)

42. Palla, G., Barabasi, A.-L., Vicsek, T.: Quantifying social group evolution. Nature **446**(7136), 664–667 (2007)

43. Parra, D., Trattner, C., Gómez, D., Hurtado, M., Wen, X., Lin, Y.-R.: Twitter in academic events: a study of temporal usage, communication, sentimental and topical patterns in 16 computer science conferences. Comput. Commun. **73**, 301–314 (2015)

44. Saganowski, S., Gliwa, B., Bródka, P., Zygmunt, A., Kazienko, P., Kozlak, J.: Predicting community evolution in social networks. Entropy **17**, 3053–3096 (2015)

45. Scholz, C., Atzmueller, M., Barrat, A., Cattuto, C., Stumme, G.: New insights and methods for predicting face-to-face contacts. In: Proceedings of the 7th International AAAI Conference on Weblogs and Social Media. AAAI Press, Palo Alto (2013)

46. Scholz, C., Atzmueller, M., Stumme, G.: On the predictability of human contacts: influence factors and the strength of stronger ties. In: Proceedings of the Fourth ASE/IEEE International Conference on Social Computing. IEEE Computer Society, Boston (2012)

47. Scholz, C., Atzmueller, M., Stumme, G.: Unsupervised and hybrid approaches for on-line RFID localization with mixed context knowledge. In: Andreasen, T., Christiansen, H., Cubero, J.-C., Raś, Z.W. (eds.) ISMIS 2014. LNCS, vol. 8502, pp. 244–253. Springer, Heidelberg (2014)

48. Scholz, C., Stumme, G., Doerfel, S., Hotho, A., Atzmueller, M.: Resource-aware on-line RFID localization using proximity data. In: Gunopulos, D., Hofmann, T., Malerba, D., Vazirgiannis, M. (eds.) ECML PKDD 2011, Part III. LNCS, vol. 6913, pp. 129–144. Springer, Heidelberg (2011)

49. Scholz, C., Illig, J., Atzmueller, M., Stumme, G.: On the Predictability of talk attendance at academic conferences. In: Proceedings of the 25th ACM Conference on Hypertext and Social Media. ACM Press, New York (2014)

50. Staab, S.: Emergent semantics. IEEE Intell. Syst. 1, 78–86 (2002)

51. Stehlé, J., Voirin, N., Barrat, A., Cattuto, C., Isella, L., Pinton, J.-F., Quaggiotto, M., Van den Broeck, W., Régis, C., Lina, B., Vanhems, P.: High-resolution measurements of face-to-face contact patterns in a primary school. PLoS ONE 6(8), e23176 (2011)

52. Steurer, M., Trattner, C.: Predicting interactions in online social networks: an experiment in second life. In: Proceedings of the 4th International Workshop on Modeling Social Media, MSM 2013, pp. 5:1–5:8. ACM, New York (2013)

53. Subrahmanyam, K., Reich, S.M., Waechter, N., Espinoza, G.: Online and offline social networks: use of social networking sites by emerging adults. J. Appl. Dev. Psychol. 29(6), 420–433 (2008)

54. Tang, L., Liu, H., Zhang, J., Nazeri, Z.: Community evolution in dynamic multi-mode networks. In: Proceedings of the ACM SIGKDD International Conference on Knowledge Discovery and Data Mining, pp. 677–685. ACM, New York (2008)

55. Tang, L., Wang, X., Liu, H.: Community detection via heterogeneous interaction analysis. Data Min. Knowl. Discov. 25(1), 1–33 (2012)

56. Turner, J.C.: Towards a cognitive redefinition of the social group. Cah. Psychol. Cogn. 1(2), 93–118 (1981)

57. Wasserman, S., Faust, K.: Social Network Analysis: Methods and Applications. Structural analysis in the social sciences, vol. 8, 1st edn. Cambridge University Press, Cambridge (1994)

A Habit Detection Algorithm (HDA) for Discovering Recurrent Patterns in Smart Meter Time Series

Rachel Cardell-Oliver[1,2]([⊠])

[1] CRC for Water Sensitive Cities, PO Box 8000, Clayton, VIC 3800, Australia
[2] The University of Western Australia, 35 Stirling Highway, Crawley 6009, Australia
rachel.cardell-oliver@uwa.edu.au

Abstract. Conserving water is a critical problem and characterising how households in communities use water is a first step for reducing consumption. This paper introduces a method for discovering *habits* in smart water meter time series. Habits are household activities that recur in a predictable way, such as watering the garden at 6 am twice a week. Discovering habit patterns automatically is a challenging data mining task. Habit patterns are not only periodic, nor only seasonal, and they may not be frequent. Their recurrences are partial periodic patterns with a very large number of candidates. Further, the recurrences in real data are imperfect, making accurate matching of observations with proposed patterns difficult. The main contribution of this paper is an efficient, robust and accurate Habit Detection Algorithm (HDA) for discovering regular activities in smart meter time series with evaluation the performance of the algorithm and its ability to discover valuable insights from real-world data sets.

Keywords: Temporal patterns · Partial periodic patterns · Habits · Smart metering · Activity recognition

1 Introduction

Conserving water is a critical problem because the world's water requirements are projected to rise beyond sustainable water supplies by 40 % by 2030[1]. Reducing water use, by identifying and changing wasteful behaviours, is one way that water utilities and their customers can address this complex problem. Data for this task can be obtained from smart meters that provide hourly, per-household consumption readings. However, reducing consumption requires a detailed understanding of the types and timing of water use behaviours. That is, it requires discovery of new knowledge from smart meter data.

This paper introduces a method for discovering *habits* in smart meter time series. Habits are activities that recur in a predictable way, such as watering the

---

[1] http://reports.weforum.org/global-risks-2015/.

M. Atzmueller et al. (Eds.): MSM, MUSE, SenseML 2014, LNAI 9546, pp. 109–127, 2016.
DOI: 10.1007/978-3-319-29009-6_6

garden at 6am twice a week. A habit can comprise several underlying household tasks, such as 2 people showering and also running a washing machine between 7 and 8 am every week day, so long as all three tasks occur during the same hour and on a regular pattern of days. Habits are known to be an important aspect of human behaviour that determine water consumption outcomes [8,19]. Habits are a good target for water conservation because they recur frequently and their volume and recurrence pattern enables the task(s) that make up the habit to be easily recognised by householders. Examples of the types of habits we aim to discover automatically are:

> HABIT [800,950] L/h OCCURS AT TIME 05:00 ON alternate DAYS FROM
> 2 Jan 2014 to 9 Apr 2014
> HABIT [1400,1500] L/h OCCURS AT TIME 22:00 ON Mo, Th, Fr DAYS
> FROM 13 Jan 2014 to 31 Jan 2014

Habits are characterised by high magnitude of water use (e.g. >1400 L/h) and their recurrence pattern (e.g. on Mondays, Thursdays and Fridays). Recurrences are anchored at a particular time of day (e.g. 05:00 or 22:00) and recur for a period of months or weeks. Each habit recurs with a recognisable pattern such as, alternate days or three times a week on Mondays, Thursdays and Fridays. Habits provide useful information to help water users to recognise their high-use behaviours and so reduce potentially inefficient use, such as over-watering the garden. At a population scale, habit discovery provides utilities with the knowledge needed for planning and management of water efficiency.

Discovering habit patterns automatically is a challenging data mining task. Habit patterns represent calendar-based activities, but they include complexities that are not addressed by existing temporal pattern discovery methods. Habit patterns are not only periodic [13], nor only seasonal [16], and they may not be frequent [10]. Habit recurrences are described by higher order regular expressions [15,17,22] with very large numbers of possible candidates. Further, the recurrences in real data are imperfect, which makes accurate matching of observations with proposed patterns difficult.

In order to address these challenges, this paper introduces a practical habit detection algorithm (HDA) for discovering regular activities in a consumption time series. The algorithm has two phases, which are illustrated in Fig. 1. First observations from the same hour of the day are clustered into groups with similar consumption magnitude (Fig. 1 left). These clusters are found at 7 am, 8 am and 7 pm. Second, HDA generates and tests the best matching temporal recurrence pattern of days for each cluster (Fig. 1 right). In this case, only one of the four candidate clusters is sufficiently regular to qualify as a habit. The days of the week when the use occurs are labelled (2,4,6,1,3,5,7,2,4,6,...) which is every second day.

The main contributions of this paper are: a model for habits using (only) features from hourly smart meter time series; an algorithm, HDA, for discovering habits; and an evaluation of HDA's handling of uncertainties in real-world data, its run time, and its real-world benefits for water utilities and their customers. The paper is organised as follows. Section 2 reviews existing methods that apply

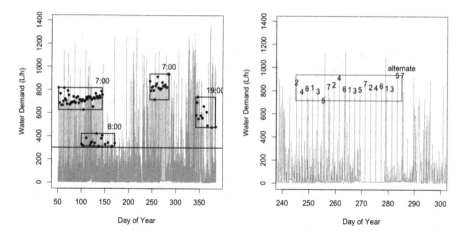

Fig. 1. Phases of the Habit Discovery Algorithm: (1) find candidate clusters (left) then (2) discover recurrence patterns (right)

to this problem. Definitions are presented in Sect. 3. An efficient algorithm for discovering habits in smart meter time series is proposed in Sect. 4. Data from real-world smart water metering trials are used to demonstrate the feasibility of the new approach in Sect. 5 which evaluates the performance of the algorithm and its ability to discover valuable insights.

2 Background and Related Work

2.1 Data Analytics for Water Consumption

There are several existing approaches for discovering knowledge from smart water meter data. Cole et al. [7] argue that a simple volume threshold (e.g. 300 L/h or 600 L/h) distinguishes between demand that corresponds to indoor use (e.g. showering) and outdoor use (e.g. garden watering). Kim et al. [12] investigated fine grained electricity and water traces over several months from a single family house. They use a matched filter algorithm to identify thresholds that disaggregate high-magnitude observations, such as garden watering, from background consumption. Cardell-Oliver proposed *water use signature patterns* from hourly time series [3]. Water use signatures can describe leaks, high magnitude ad hoc peaks, and regular high magnitude events. The first algorithm for habit discovery was proposed in [4]. This paper extends that work with a significantly more efficient algorithm and more detailed evaluation of its performance. Wang et al. [21] introduces a related algorithm that discovers frequent patterns that are subsequences (consecutive hours) of water use. The sequence patterns discovered in that work are frequent but not necessarily regular.

Several machine learning methods have been proposed for disaggregating water use events into end-uses such as showering or flushing a toilet. The most common approach uses supervised learning, in which an expert is required to label training data for each household Tools such as TraceWizard [2,18] or Hydrosense [9] can support the labelling process, but manual labelling is still too time consuming to be used on a large scale. Unsupervised learning methods do not require training data. FixtureFinder [20] uses unsupervised learning to automatically infer fixtures (e.g. taps) in the home by fusing smart water meter data with other sensors such as electricity or home security. Chen et al. [5] proposed a novel statistical framework to disaggregate water use events from coarse-grain smart meter data and tested it on simulated data with periods of 15 min to 3 h. However, acceptably accurate activity recognition is only possible given measurement frequency of (sub-)seconds [2,18,20] or a few minutes [5]. So the amount of data that must be collected and stored is impractical for town or city scale monitoring.

HDA uses a different approach. Two features are used to characterise patterns: hourly volume and temporal recurrence. Combinations of these features trigger a householder's identification of the specific activities underlying their habits. For example, when shown the habit, HABIT [800,950] L/h OCCURS AT TIME 05:00 ON alternate DAYS FROM 2 Jan 2014 to 9 Apr 2014 from their recent water use history, a householder is likely to be able to identify what they were doing at that time. For example, this habit is likely to correspond to the operation of an automatic garden watering system, maybe combined with some other lower volume indoor activities at 5 am. HDA does not attempt to identify the latent factors, but only to recognise groups of observations with similar volumes and strong temporal constraints on their recurrence.

2.2   Calendar Patterns

Previous studies have shown that machine learning can reveal valuable knowledge about water use activities. However, none of these approaches provide specific information about the regularity or recurrence of activities. This section reviews related work on discovering periodic patterns in time series data.

Partial periodic patterns (PPPs) were introduced in 1999 [11]. A PPP is a template for pattern matching. It is a tuple containing literals and don't-care symbols: *. For example, the PPP a*c matches the sequences abc and aac but not abb. The length of the tuple is the period of the pattern, in this case 3. Partial patterns are distinguished from full periodic patterns in that PPPs have one or more don't-care positions, while full patterns have a literal in every position. Han et al. [11] present an efficient algorithm for discovering partial patterns that first finds all patterns with only one literal, and then combines these to generate a maximum matching pattern, from which sub-patterns can be checked to find all frequently occurring partial patterns for a given time series.

Yang et al. [22] introduce higher-order partial patterns. These extend the model of PPPs by allowing patterns themselves to appear as tuple elements in a pattern. For example, ((a*c):[1,30]; *:[31:40]; (aa):[41:50]) matches a sequence in which the period-3 pattern a*c occurs 10 times between 1 and 30, there is a break of length 10, and then the period-2 pattern **aa** occurs 5 times between times 41 and 50.

The matching criteria in existing PPP algorithms is binary: either an observed sequence matches a pattern or it does not. There is no allowance for near miss matches. Li et al. [13,14] propose a calendar method where pattern match tuples are regular expressions on calendar schemas such as (year, month, day). The user specifies particular patterns of interest, and inclusiveness and exclusiveness metrics are used to select the best matching pattern. These measures take into account the number of false positives or false negatives within patterns that contribute to the match.

HDA discovers partial periodic patterns using a generate and test approach. An alternative approach is to learn recurrences directly from the data. Grammar based encoding is a promising technique for this task. Sequitur [17] is an algorithm for learning a regular (context free) grammar from a symbolic sequence. In a single pass through the data, the algorithm learns rules that summarise the data. Processing the sequence abcabcabd would generate the encoding CAd with grammar rules A → ab, B → Ac, C → BB (which can be simplified to, A → ab, C → AcAc). Learning regular grammars is a promising approach for future work on describing cyclic behaviours. However, there are some challenges to be addressed. Firstly, Sequitur is a greedy algorithm that may make sub-optimal decisions about the grammar rules. This can result in difficult to understand encodings. For recurring household activities, some form of run length encoding would be useful. For example, (abc)^{10} meaning that pattern abc occurs 10 times. Secondly, like the PPP algorithm, Sequitur uses exact matching: an observed sequence either matches an existing rule or it does not. In order to deal with noisy real-world times series, we need to make finer distinctions about matches, similar to the information theoretic measures used in HDA. These open questions for learning grammars for habits are outside the scope of this paper, and are left for future research.

3   Habit Discovery Definitions

This section introduces notation and definitions for regular, high magnitude habits in smart meter time series. Smart water meter time series measure the water consumption of a household or business. In general, a smart meter time series can represent any measure of consumption such as gas or electricity. For simplicity, the term "volume" will be used throughout this paper.

Definition 1. *A* **smart (water) meter time series** $T = (t_1, t_2, \cdots, t_n)$ *is a sequence of n consecutive hourly water consumption observations for an individual property.*

Each observation t has three attributes: $day(t)$ and $hour(t)$ identify the day of the year(s) $(1\ldots365\ldots)$ and hour of day $(0\ldots23)$ of the observation and $vol(t)$ is a real number representing the volume in litres of water consumed during one hour. A unique timestamp for each observation is given by $timestamp(t) = day(t) * 24 + hour(t)$.

For a set of observations, S, we write $day(S)$ and $hour(S)$ for the sets obtained by projecting S onto its day or hour attributes. The projection, $vol(S)$ of S onto volume attributes is a bag (rather than a set) because the same volume may occur more than once in the records.

3.1   Candidate Selections

Definition 2. *A **candidate selection** is a set of observations constrained by volume, day and hour parameters. A candidate selection from time series T is defined by the function $CS(T, v_1, v_2, d_1, d_2, h) =$*

$$\{t \in T \mid vol(t) \in [v_1, v_2] \ \wedge \ day(t) \in [d_1, d_2] \ \wedge \ hour(t) = h\}$$

The volume and time bounds are tight for candidate selection S. That is, $v_1 = min(vol(S))$, $v_2 = max(vol(S))$, $d_1 = min(day(S))$ and $d_2 = max(day(S))$.

The tight parameter bounds ensure that the intensional definition of a candidate selection by parameter list $(T, v_1, v_2, d_1, d_2, h)$ uniquely defines a set of observations from T, and so we can use either notation: the intensional parameter list or the extensional set of observations. The number of observations in a candidate selection set S is $size(S)$, and its centroid (mean) volume is $centroid(S) = \Sigma_{t \in S} vol(t)/size(S)$.

3.2   Calendar Patterns

Candidate selections are sets of observations that are cohesive in their spread of days, hour of day and volume of use. Habits are candidate selections that also have a regular temporal recurrence. This section defines the types of regular calendar pattern recognised by HDA. Each HDA pattern is a partial periodic pattern [11] on a given time interval [22]. A typical user has several HDA habits. HDA can detects two classes of calendar patterns.

Definition 3. *A **day pattern**, $R(d_0, k, m)$ is an arithmetic sequence of m days (positive integers) starting at day d_0 and recurring every k days. $R(d_0, k, m)$ denotes the sequence generated by $d_i = d_0 + i \times k$ for $i \in 0 \ldots m - 1$.*

For example, every day starting at day 15 and recurring 5 times is written $R(15, 1, 5)$. This pattern comprises 5 days: *15, 16, 17, 18, 19*. The alternate day pattern $R(15, 2, 8)$ comprises a sequence of 8 (odd) days: *15, 17, 19, 21, 23, 25, 27, 29*. Every Monday for 4 weeks, where day 10 is a Monday, is written $R(10, 7, 4)$ which denotes the days *10, 17, 24, 31*.

Activities that occur regularly on two or more days of the week are called calendar patterns.

Definition 4. *A **calendar pattern**, $R^*(d_0, W, m)$ occurs over m weeks on selected days of the week $W \subseteq 1..7$ where $nextMonday(d_0)$ is the first Monday following d_0. The pattern is constructed as the union of its contributing single-day patterns.*

$$R^*(d_0, W, m) = \bigcup_{w \in W} R(nextMonday(d_0) + w - 1, 7, m)$$

For example, $R^*(8, \{2, 4\}, 4)$ denotes *10, 13, 17, 20, 24, 27, 31, 34* since $nextMonday(8)=10$ then every Tuesday (2) and Thursday (4) for 4 weeks. A day pattern is a restricted type of calendar pattern, and so we use the general term, calendar pattern, for both.

3.3   Habits

Habits are candidate selections that satisfy strong constraints on their magnitude and recurrence. The constraints are expressed using five parameters:

1. *min_sup*: minimum number of hourly observations
2. *min_vol*: a lower bound for all volumes in the habit
3. *vol_range*: the maximum volume range
4. *max_dis*: the maximum gap between successive days
5. *min_match*: a minimum F-measure for the correspondence with a regular recurrence pattern of days chosen from a list of possible regular expressions.

Definition 5. *A **habit** comprises a candidate selection and calendar pattern where the following constraints are satisfied for constants, min_sup, min_vol, vol_range, max_dis, and min_match:*

Property	Constraint
Candidate selection	$S = CS(T, v_1, v_2, d_1, d_2, h)$
Frequency	$size(S) > min\_sup$
Significance	$v_1 = min(vol(S)) > min\_vol$
Volume coherence	$v_2 - v_1 < vol\_range$
Temporal persistence	$\forall t_i \in S.\ time(t_i) \neq max(time(S)) \Rightarrow \exists\, t_j \in S.\ 0 < (t_j - t_i) < max\_dis$
Calendar pattern	$cp = R^*(d_0, W, m)$
Regular recurrence	$F\_measure(days(S), cp) < min\_match$

The habit corresponding to candidate selection $CS(T, v_1, v_2, d_1, d_2, h)$ and calendar pattern cp can also be written in a more human readable form as:

HABIT $[v_1, v_2]$ L/h OCCURS AT TIME h ON cp DAYS FROM d_1 TO d_2

For example,

HABIT [400,600] L/h OCCURS AT TIME 22:00 ON Mo, Th, Fr DAYS
FROM 13 Jan 2014 to 31 Jan 2014

Algorithm 1. Add an observation to a candidate selection $CanUpdate$

    Input   : S a candidate selection;
                 t_i the next observation from smart meter time series T;
    Output   : S^* an updated candidate selection

1   Let $S = CS(T, vol_1, vol_2, day_1, day_2, h)$
2   **if** $h \neq hour(t_i)$ **then**
3 | **return** Error(Hour of day fields must match)
4   $vv_1 = min(vol_1, vol(t_i))$
5   $vv_2 = max(vol_2, vol(t_i))$
6   $dd_1 = min(day_1, day(t_1))$
7   $dd_2 = max(day_2, day(t_2))$
8   **return** $CS(T, vv_1, vv_2, dd_1, dd_2, h)$

---

4   Habit Discovery Algorithm

The data mining problem of habit discovery addressed in this paper is as follows.

Given: time series T and constants min_sup, min_vol, vol_range, max_dis, and min_match

Identify: all candidate selections $S = CS(T, v_1, v_2, d_1, d_2, h)$ and calendar patterns $cp = R^*(d_0, W, m)$ where (S, cp) is a habit, and

Report: a list of habits, each specified by its candidate selection parameters (v_1, v_2, d_1, d_2, h) and its calendar pattern cp.

HDA has two phases: sequential clustering of observations into candidate selections and generating and testing the best matching recurrence for each cluster. The habit discovery problem can not be solved exactly because the number of possible selections in a time series is infeasibly large. So the HDA algorithm uses heuristics to search efficiently for habits.

4.1   Identify Candidate Selections

Candidate selections are grown incrementally, one observation at a time. Algorithm 1 gives the function $CanUpdate$ for this task. This function fails if the hour of the new observation is not the same as those of the existing candidate selection. Otherwise it returns the tuple of bounds that describes the incremented candidate selection.

Each new observation is added to its nearest candidate selection. Algorithm 2 shows the function $CanDist$ which uses four cases for determining this distance. The first two cases are negative matches: the habit conditions of stability (same hour of day lines 2–3) or temporal persistence (lines 4–5) are *not* satisfied. The last two cases are positives when either the observation already lies within the candidate's volume bounds (reported as distance 0, lines 6–7) or, if the observation is outside the current bounds of the candidate, then the difference between the observation volume and the candidate's centre is returned (lines 8–9).

Algorithm 2. Distance Function for Candidate Selection $CanDist$

Input   : t_i observation from T, S a candidate selection
Parameters: min_vol volume lower bound;
              vol_range maximum volume range;
              max_dis maximum gap between days;
Output   : return distance measure from t_i to S

1   Let $S = CS(vol_1, vol_2, day_1, day_2, h)$
2   **if** $hour(t_i) \neq h$ **then**
3     $\lfloor$ **return** ∞ ;   // wrong hour of day
4   **if** $time(t_i) - day_2 > max\_dis$ **then**
5     $\lfloor$ **return** ∞ ;   // out of time range
6   **if** $vol(t_1) \in [vol_1, vol_2]$ **then**
7     $|$ **return** 0 ;   // within existing time range
8   **else**
9     $\lfloor$ **return** $\mid vol(t_i) - centroid(S) \mid$; // return distance from volume group
       centre

4.2   Identify Recurrences

After generating candidate selections the HDA discovery algorithm filters the candidates, retaining only those with a strong recurrence patterns. The function $CanMatch$ generates and tests potential patterns and returns the best matching pattern and its F-measure. This version of the algorithm tests three possible periods for patterns: every-day, 2-day and 7-day, chosen to cover the expected recurrences for high magnitude activities such as garden watering. This section explains the implementation of the $CanMatch$ function given in Algorithm 3.

Matching Recurrences. For the noisy patterns of days encountered in real-world smart meter traces, there is more than one possible recurrence pattern. To deal with this situation we need a measure for the closeness of the match between an observed sequence of days and a given day pattern or calendar pattern. The F-measure score from information retrieval is used for this purpose. Suppose we have a candidate selection with day range 20 to 50. where the actual days in the selection are: $A = \{20, 22, 24, \mathbf{25}, 28, 30, 32, 38, 40, 42, 44, 46, 48, 50)\}$. The pattern of alternate (even) days, for the same period, would be $E = \{20, 22, 24, \mathbf{26}, 28, 30, 32, \mathbf{34}, \mathbf{36}, 38, 40, 42, 44, 46, 48, 50\}$ from the regular expression $(10)^*$. Days in bold font indicate differences between E and A. These differences are captured by standard information retrieval metrics: true positives $TP = E \cap A$; false positives $FP = A \backslash E$, in this case 25, the value that was not expected but was observed; and false negatives $FN = E \backslash A$, in this case 26,34,36, the values that were expected but were not observed. Now we have, $precision = size(TP)/size(TP \cup FP)$ and $recall = size(TP)/size(TP \cup FN)$ and $F\_measure = (2 \times precision \times recall)/(precision + recall)$. For the example, precision is $13/(13+2)$ and recall is $13/(13+3)$ giving an F-measure of 0.84.

Algorithm 3. Find best matching calendar pattern *CanMatch*

Input   : Candidate selection $S = CS(T, v_1, v_2, d_1, d_2, h)$;
Output   : best pattern *best_cp* and pattern match ratio *best_r*
// Generate possible calendar patterns
1 *scope* $= d_2 - d_1 + 1$
2 *everyday* $= R(d_1, 1, scope)$
3 *alternate1* $= R(d_1, 2, scope/2)$
4 *alternate2* $= R(d_1 + 1, 2, scope/2)$
   // Permute of days the week in their descending frequency in S
5 Let *most_common_days* $= \langle w_1, w_2, w_3, \ldots w_7 \rangle$
6 **for** $k = 1$ **to** 7 **do**
7 $\quad$ $pat\{w_1, \ldots, w_k\} = R^*(d_1, \{w_1, \ldots, w_k\}, scope/7)$

   // Select the pattern with best match to days of S
8 *Patterns* $= \langle$ *everyday, alternate1, alternate2, pat*$\{w_1\}, \ldots,$ *pat*$\{w_1, \ldots, w_7\} \rangle$
9 *best_r* $= 0$;   *best_cp* $=$ *none*
10 **for** $cp \in Patterns$ **do**
11 $\quad$ $r = F\_measure(S, cp)$
12 $\quad$ **if** $r > best\_r$ **then**
13 $\quad\quad$ *best_r* $= r$
14 $\quad\quad$ *best_cp* $= cp$

15 **return** $(best\_cp, best\_r)$

Generating Recurrences. For water metering applications, *CanMatch* checks three types of calendar patterns:

– every day in the scope of the candidate selection;
– alternate days, either odd or even;
– most relevant combinations of days of the week, e.g. Mon, Thu, Sat.

Suppose candidate selection S has 132 days corresponding to days of the week: Mo, We and Fr occurring 46, 43 and 38 times, respectively and also Tu (2 times), Th (1) and Su (2). The pattern matches for this example are shown in the following, with the best matching pattern is $\{Mo, We, Fr\}$. *CanMatch* for this S returns $(pat\{Mo, We, Fr\}, 0.92)$.

4.3   HDA Algorithm

Algorithm 4 summarises the complete HDA algorithm. The first phase (lines 1 to 12) uses sequential clustering to identify candidate selections. This simple algorithm has two major advantages: it is linear on $n \times q$ for input size n and number of candidate selections q; and it does not need the number of candidate selections to be decided in advance. The algorithm can also be run in stream mode, processing smart meter observations as they become available. The second phase (lines 13 to 20) filters the discovered candidates selections by their size and the strength of their recurrence pattern.

Pattern used for matching	F-measure
everyday	0.57
alternate1	0.41
alternate2	0.47
$pat\{Mo\}$	0.51
$pat\{Mo, We\}$	0.78
$pat\{Mo, We, Fr\}$	0.92
$pat\{Mo, Tu, We, Fr\}$	0.80
$pat\{Mo, Tu, We, Fr, Su\}$	0.71
$pat\{Mo, Tu, We, Th, Fr, Su\}$	0.63

5   Experimental Evaluation

This section evaluates four aspects of the HDA algorithm: parameter settings, limitations of the algorithm, running time, and its utility for the water industry.

5.1   Datasets

Real-world smart meter readings from two contrasting populations are used to evaluate the HDA algorithm.

Karratha is a town in the north-west of Western Australia, about 1200 Km north of the capital, Perth. Its population has grown from 17,000 in the 2011 census to an estimated 25,000 in 2013, and forecast to reach 41,000 by 2036. The climate is semi-arid. In the study period of 2014 the maximum day reached 43.3 degrees Celsius and the minimum recorded was 8.4 degrees. Annual daily averages are 33.1 degrees maximum and 20.7 minimum. There were 18 days of rain in 2014 giving annual rainfall of 221.0 mm.

Kalgoorlie-Boulder is a city of 31,000 people in the Eastern Goldfields of Western Australia, 600 km east of the capital, Perth. The climate is semi-arid. In the study period of 2014 the maximum day reached 43.2 degrees Celsius and the minimum recorded was –3.4 degrees. Annual daily averages are 26.2 degrees maximum and 12.7 minimum. There were approximately 48 days of rain during this period giving rainfall of 359.2 mm which is higher than the annual average of 264 mm.

Each dataset comprises readings for a full year for 500 meters. The datasets include smart meter time series for different land uses including houses, duplex, triplex and quadruplex units, home units and flats. Houses comprise 92 % of the sample. For consistency, only houses are used for our analysis and properties with annual consumption below 25 L/day are also removed from the sample. After this pre-processing there are 466 time series for Karratha and 440 for Kalgoorlie.

5.2   Parameter Values

Setting parameters for knowledge discovery algorithms is difficult because optimal parameter settings differ for different domains and business applications,

---

Algorithm 4. Habit Detection Algorithm (HDA)

Input   : T a smart meter time series of length n
Parameters: min_sup minimum support;
            min_vol lowest supported volume vol_range volume range;
            $max_d is$ maximum gap between days;
            min_match minimum F_measure for pattern match
Output   : H set of discovered habits

```
   // Initialisation
1  Set m = 0 and initialise candidate selection for each hour as C_h = ∅
   // Identify candidate selections by adding each t_i ∈ T
2  for i = 1 to n do
3  |   if vol(t_i) > min_vol then
4  |   |   h = hour(t_i)
5  |   |   d = min(CanDist(t_i, S)) × 2 for S ∈ C_h or d = ∞ when C_h = ∅
6  |   |   if d ≤ vol_range/2 then
       |   |   // grow an existing candidate selection
7  |   |   |   S_new = CanUpdate(S, t_i)
8  |   |   |   C_h = C_h\S ;                      // remove the old candidate
9  |   |   |   C_h = C_h ∪ S_new ;                // replace with the new
10 |   |   else
       |   |   // start growing a new candidate selection
11 |   |   |   S_new = (T, hour(t_i), vol(t_i), vol(t_i), day(t_i), day(t_i))
12 |   |   |   C_h = C_h ∪ {S_new}

   // Select all valid candidate selections
13 H = ∅
14 for h = 0 to 23 do
15 |   for each S ∈ C_h do
16 |   |   if size(S) ≥ min_sup then
17 |   |   |   (p, r) = CanMatch(S)
18 |   |   |   if r > min_match then
19 |   |   |   |   H = H ∪ {(S, p)}

20 return H
```

---

and small changes in the parameters can have a large effect on the results. On one hand, the parameter settings are a subjective choice based on the domain knowledge of experts. HDA's parameters are chosen to allow for uncertainty and noise in the input data. On the other hand, parameter choices can have unexpected effects on the results. In particular, we wish to avoid parameter choices where a small change in the parameter has a large effect on the outcome. This section assesses the choice of parameters from these two perspectives. Except where otherwise stated, the parameters used for evaluation of HDA are

1. $min\_sup = 7$ days.
2. $min\_vol = 300$ L/h.

3. $vol\_range = 300$ L.
4. $max\_dis = 10$ days.
5. $min\_match = 0.8$.

Allowing for Uncertainties. The water meters used in this study report the volumes of water consumed each hour. There are several sources of uncertainty be considered when identifying habits. First, there are *errors of measurement* caused by rounding down the reported hourly volume to the nearest Litre, uncertainty around the time of the hourly reading, and the limited accuracy of flow meters, which is typically around $\pm 2\,\%$. These sources of noise could thus account for differences for up to 40 L/h in the reported values for the same 1000 L/h activity.

Second, the hourly volumes can include several *sequential or concurrent human activities* such as taking showers and washing up breakfast while an automatic garden watering system is running. This source of uncertainty could account for 300 L/h difference between repeated recurrences of the same outdoor activities (such as garden watering). Cole et al. [6] analyse indoor and outdoor water use and argue that use above 300 L/h most likely indicates outdoor activity. The HDA parameters $vol\_range = 300$ L/h and $min\_vol = 300$ L/h allow for these types of uncertainty.

A third source of noise is that caused by *irregularities in human behaviour*. For example, householders may turn off their watering systems for one or two days when there has been rain or run an extra day during a heat wave. The parameters $min\_sup = 7$ days, $max\_dis = 10$ days and $min\_match = 0.8$ allow for these types of uncertainty.

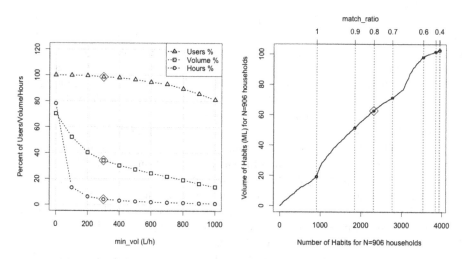

Fig. 2. Effect of (left) min_vol on the volume of water and number of hours selected from full data sets and (right) different match ratios for recurrence. Default values of the parameters are indicated by diamonds.

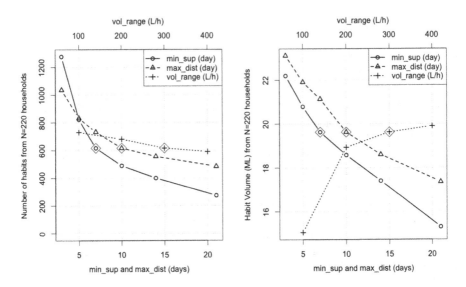

Fig. 3. Effect of *min_sup*, *max_dist* and *vol_range* on number of discovered habits (left) and the total volume of those habits for 220 households. Default values of the parameters are indicated by diamonds.

Stability. Ideally parameter choices should be stable in that small changes in the parameter value should not have large effects on the results of HDA. For analysing stability we consider three properties of HDA's outputs: the *number of habits* in the list, the *total volume* of water consumed for these habits, and the *total hours* covered by the habits.

Figure 2 (left) shows the effect of the choice of *min_vol* in the first step of HDA, in which only hourly volumes greater than *min_vol* are selected for further analysis. This step affects the performance of the algorithm by reducing the number of hours of water use to be analysed in the later stages of the algorithm. It can be seen that choosing *min_vol* = 300 L/h (shown by the diamonds on each series) restricts the number of hours for analysis to 4 %, while still covering 34 % of all water demand and 98 % of households.

Figure 2 (right) shows the effect of increasing the *match_ratio*. The graph line is the cumulative sum of the number of habits and volume of those habits, ordered by the *match_ratio* for each combination. The vertical line for *match_ratio* = 1 shows that accepting only perfect pattern matches returns 20 % of all volumes and 25 % of all habits. The vertical line for the default *match_ratio* = 0.8 shows this setting returns around 60 % of habits and their volume. Furthermore, the gradient of change in this region is gradual, suggesting 0.8 is a reasonably stable parameter choice.

Figure 3 shows the effect of varying the parameters *min_sup*, *max_dist*, *vol_range* on the final results of HDA. This figure shows the number of habits (left) and the total volume of water used in those habits (right) for each parameter choice for a population of 220 households from a single suburb of Karratha.

One parameter was varied at a time, with the other two held at their default values. The gradual slope of the graphs around each default parameter (indicated by a diamond) indicates that the default parameter choices are reasonably stable for both number of habits and their overall volume.

5.3   Limitations

Because HDA uses heuristics to search for habits, it is not guaranteed to find all habits in a smart meter time series. This section identifies three scenarios for missed habits that were discovered by inspection in the case study time series.

Unusual Periodicity. Habits with unusual periods are not detected by HDA. For example, the candidate selection on days *(177, 180, 183, 186, ..., 258)* is incorrectly classified by HDA as HABIT [485,671] L/h OCCURS AT TIME 22:00 ON every DAYS (F1 = 0.51) FROM day 177 to day 258. Since there is no check for patterns with period 3. Because this match ratio is less than 0.8, this habit is rejected. In this case HDA does not correctly identify the 3-day periodic pattern since only expected recurrence periods of 1, 2 or 7 days were tested. This problem can be solved by programming HDA to check for unusual periods (such as 3 to 6), at the expense of longer run times.

Sudden Pattern Changes. Habits with a sudden change of recurrence pattern are not correctly classified by HDA since there is no gap of *max_dis* days between the patterns. For example, the sequence of days with recurrence 1,3,5,7 for 32 weeks, followed immediately by recurrence 1,2,4,6 for 12 weeks was incorrectly classified as HABIT [384,544] L/h OCCURS AT TIME 6:00 ON {1,3,5,7} DAYS (F1 = 0.8) FROM day 45 to day 366   That is, the change of recurrence pattern was not identified. This problem could be solved by matching candidate selections to recurrence patterns in stream mode (e.g. one week at a time) rather than matching the whole selection to a single recurrence pattern. Grammar learning algorithms for this task, such as Sequitur [17], will be investigated in future work.

Crossed Volume Boundaries. Habits that cross volume boundaries may not be discovered by HDA. For example, a habit with range [600,700] may be misclassified if bounds [500,650] and [650,800] are selected during the candidate selection phase. In that case, the days used for pattern matching will not include some observations of the habit and may include spurious observations, leading to a low match ratio, and rejection of the candidate selections. This is a known problem for the sequential clustering algorithm used by HDA to identified candidate selections. A possible solution, to be investigated in future work, is to use limited back-tracking to delay the assignment of a distant point to a new cluster. This approach reduces the possibility of mis-aligning cluster centroids, and so results in more coherent clusters.

5.4   Run Time

This section evaluates the efficiency of the HDA algorithm in terms of its run time for different case study datasets. We also compared the run time of the improved HDA algorithm presented in this paper (HDA) with the original algorithm (HDAv0) presented at SenseML [4]. Both versions of HDA are implemented in the R system[2]. The R code source files for the HDA algorithm is 320 commented lines of code. The code for both versions was run on a MacBook Air 1.3 GHz Intel Core i5 with 4 GB memory.

Table 1. Runtime and Space for HDA

Algorithm	Population (1000 hh)	Scope (days/hh)	Runtime (minutes)
HDAv0	0.1	145	0.61
HDAv0	0.5	145	2.11
HDAv0	0.1	368	3.00
HDAv0	0.5	368	14.40
HDA	1	365	1.27
HDA	15	365	20.32

The current version of HDA takes just over a minute to discover the habits of approximately 1000 customers with 1 year of data. The final row of Table 1 shows that run time is just over 24 min for a town-size population of 15,000 customers over 350 days. This data was generated from sampled case study meters. This demonstrates that HDA is sufficiently fast for real-world business applications. The original HDAv0 algorithm (rows 1 to 4 of Table 1) is significantly slower. The main difference between HDAv0 and HDA is in the clustering phase of the algorithm. The earlier version used an agglomerative clustering algorithm to determine the volume boundaries, and applied the clustering step before separating the data into individual hours of the day. The current algorithm uses sequential clustering and applies the clustering step after the data has been partitioned by hours of the day.

5.5   Practical Applications

This section illustrates how water utilities and customers can benefit from habit discovery. Table 2 summarises some significant business insights that were discovered using HDA. All results use the default parameters given in Sect. 5.2.

Volume Significance of Habits. General advertising campaigns can be used to educate and inform customers about potentially wasteful activities. For the greatest effect, such advertising campaigns should be focussed on activities that account for a significant volume of water use and are prevalent amongst users.

---

[2] http://www.r-project.org/.

Table 2. HDA insights for business applications

Category	Karratha (N = 466)	Kalgoorlie (N = 440)
Number of habits identified by HDA	1338	974
Number of users with at least one habit	294 users	196 users
Volume contribution of habits to all use (annual)	20 %	13 %
Proportion of intensive habits (>1000 L/h)	42 %	53 %
Proportion of habits exceeding roster frequency	71 %	19 %
Number of users exceeding roster frequency	184 users	70 users
Peak hour of day for all consumption	6 am	7 pm
Top 3 peak hours for habits	6, 7, 5 am	6, 7, 5 am

Rows 1 to 3 of Table 2 show that the proportion of users, and the volume of water consumed by habits is significantly higher in Karratha than Kalgoorlie. If (as we believe) a high proportion of habits discovered represents automatic garden watering events, then Karratha is a stronger target than Kalgoorlie for campaigns on more efficient garden watering.

Roster Compliance and Over-Watering. In Western Australia, customers are obliged to comply with garden watering rosters. Watering days are allocated based on the climate zone of the town and the street number of the property. Identifying non-compliant watering activity can be used in one-to-one interventions with customers to reduce their water use, as well as informing policy for watering restrictions. Demand of more than 1000 L/h may be over-watering. The roster for Karratha provides for alternate day watering on either odd or even days of the month, which broadly equates to 3.5 days per week. Thus Karratha habits that occur 4 or more times may be garden watering activity that is not compliant with the rostered frequency. For Kalgoorlie the watering roster is two times a week, so habits 3 or more times are week are the target. Rows 4 to 6 of Table 2 shows that possible over-watering and roster non-compliance is a strong target for water savings in Karratha but less so in Kalgoorlie.

Peak Demand. Understanding peak demand is an important problem in the water industry since reducing peaks can lower both operating (e.g., pumping) and infrastructure (e.g., new plants) costs [1,7]. For both populations, rows 7 to 8 of Table 2 show that peak habit demand is concentrated between 5 and 7 am. For Karratha peak habits coincide with the overall daily peak, but not in Kalgoorlie. An interesting business insight is that reducing habit water use in Karratha could not only save water overall, but could also significantly reduce the daily peak. However, the daily peak in demand in Kalgoorlie is not linked to habits.

6   Conclusions and Future Work

Habits are regular, high-magnitude patterns of water consumption. This paper presents an algorithm, HDA, for detecting habits in water meter time series. HDA allows for uncertainties in real-world data. It is also sufficiently fast to be used on data sets for town or city populations. Detecting habits with HDA generates significant insights for industry. The profiles of habits for different populations can be used to design evidence-based water efficiency programs. Habit rules are understandable by humans, and so can be used by individuals to identify ways to reduce their water bills. The HDA algorithm can identify and explain the extent of regular, high-magnitude habits and the proportion of that use that is potentially inefficient or does not comply with restrictions.

In future work, HDA could be generalised to identify more types of patterns with greater robustness. Habit discovery could also be investigated in other domains such as medical or manufacturing time series.

Acknowledgments. This research is funded by the Cooperative Research Centre for Water Sensitive Cities under project C5.1. The author would like to thank H. Gigney, S. Atkinson, G. Peach and R. Pickering at the Water Corporation of Western Australia for the smart meter datasets and for their advice on interpreting them. Thanks also to Eneldo Loza Mencia, Jin Wang and the anonymous reviewers for comments that greatly improved the manuscript. This research has been approved by the Human Research Ethics Office (HREO) of the University of Western Australia (RA/4/1/6253).

References

1. Beal, C., Stewart, R.: Identifying residential water end uses underpinning peak day and peak hour demand. J. Water Resour. Plann. Manag. **140**(7), 04014008 (2014). http://dx.doi.org/10.1061/(ASCE)WR.1943-5452.0000357
2. Beal, C.D., Stewart, R.A., Fielding, K.: A novel mixed method smart metering approach to reconciling differences between perceived and actual residential end use water consumption. J. Cleaner Prod. **60**, 116–128 (2011). http://www.sciencedirect.com/science/article/pii/S0959652611003386
3. Cardell-Oliver, R.: Water use signature patterns for analyzing household consumption using medium resolution meter data. Water Resour. Res. 1–11 (2013). http://onlinelibrary.wiley.com//10.1002/2013WR014458/abstract
4. Cardell-Oliver, R.: A habit discovery algorithm for mining temporal recurrence patterns in metered consumption data. In: 1st International Workshop on Machine Learning for Urban Sensor Data (SenseML), 15 September 2014. https://www.tk.informatik.tu-darmstadt.de/en/senseml-2014/
5. Chen, F., Dai, J., Wang, B., Sahu, S., Naphade, M., Lu, C.T.: Activity analysis based on low sample rate smart meters. In: Proceedings of the 17th ACM SIGKDD International Conference on Knowledge Discovery and Data Mining, KDD 2011, pp. 240–248. ACM, New York (2011)
6. Cole, G., O'Halloran, K., Stewart, R.A.: Time of use tariffs: implications for water efficiency. Water Sci. Technol.: Water Supply **12**(1), 90–100 (2012)

7. Cole, G., Stewart, R.A.: Smart meter enabled disaggregation of urban peak water demand: precursor to effective urban water planning. Urban Water J. **10**(3), 174–194 (2013)

8. Erickson, T., Podlaseck, M., Sahu, S., Dai, J.D., Chao, T., Naphade, M.: The dubuque water portal: evaluation of the uptake, use and impact of residential water consumption feedback. In: Proceedings of the SIGCHI Conference on Human Factors in Computing Systems, CHI 2012, pp. 675–684. ACM, New York (2012). http://www.acm.org/10.1145/2207676.2207772

9. Froehlich, J., Larson, E., Saba, E., Fogarty, J., Campbell, T., Atlas, L., Patel, S.: A longitudinal study of pressure sensing to infer real-world water usage events in the home. In: Lyons, K., Hightower, J., Huang, E.M. (eds.) Pervasive 2011. LNCS, vol. 6696, pp. 50–69. Springer, Heidelberg (2011)

10. Han, J., Cheng, H., Xin, D., Yan, X.: Frequent pattern mining: current status and future directions. Data Min. Knowl. Discov. **15**(1), 55–86 (2007)

11. Han, J., Dong, G., Yin, Y.: Efficient mining of partial periodic patterns in time series database. In: Proceedings of the 15th International Conference on Data Engineering, pp. 106–115. IEEE (1999)

12. Kim, Y., Schmid, T., Srivastava, M.B., Wang, Y.: Challenges in resource monitoring for residential spaces. In: Proceedings of the First ACM Workshop on Embedded Sensing Systems for Energy-Efficiency in Buildings, pp. 1–6. ACM (2009)

13. Li, Y., Ning, P., Wang, X.S., Jajodia, S.: Discovering calendar-based temporal association rules. Data Knowl. Eng. **44**(2), 193–218 (2003)

14. Li, Y., Zhu, S., Wang, X.S., Jajodia, S.: Looking into the seeds of time: discovering temporal patterns in large transaction sets. Inf. Sci. **176**(8), 1003–1031 (2006). http://www.sciencedirect.com/science/article/pii/S0020025505000563

15. Li, Y., Lin, J.: Approximate variable-length time series motif discovery using grammar inference. In: Proceedings of the Tenth International Workshop on Multimedia Data Mining, MDMKDD 2010, pp. 10:1–10:9. ACM, New York (2010)

16. Mahanta, A.K., Mazarbhuiya, F.A., Baruah, H.K.: Finding calendar-based periodic patterns. Pattern Recogn. Lett. **29**(9), 1274–1284 (2008). http://www.sciencedirect.com/science/article/pii/S0167865508000512

17. Nevill-Manning, C.G., Witten, I.H.: Compression and explanation using hierarchical grammars. Computer J. **40**(2 and 3), 103–116 (1997). http://comjnl.oxfordjournals.org/content/40/2_and_3/103.abstract

18. Nguyen, K., Zhang, H., Stewart, R.A.: Development of an intelligent model to categorise residential water end use events. J. Hydro-environment Res. **7**(3), 182–201 (2013). http://www.sciencedirect.com/science/article/pii/S1570644313000221

19. Russell, S., Fielding, K.: Water demand management research: a psychological perspective. Water Resour. Res. **46**(5) (2010). http://www.dx.org/10.1029/2009WR008408

20. Srinivasan, V., Stankovic, J., Whitehouse, K.: Fixturefinder: discovering the existence of electrical and water fixtures. In: Proceedings of the 12th International Conference on Information Processing in Sensor Networks, IPSN 2013, pp. 115–128. ACM, New York (2013). http://www.acm.org/10.1145/2461381.2461398

21. Wang, J., Cardell-Oliver, R., Liu, W.: Efficient discovery of recurrent routine behaviours in smart meter time series by growing subsequences. In: Cao, T., Lim, E.-P., Zhou, Z.-H., Ho, T.-B., Cheung, D., Motoda, H. (eds.) PAKDD 2015. LNCS, vol. 9078, pp. 522–533. Springer, Heidelberg (2015)

22. Yang, J., Wang, W., et al.: Discovering high-order periodic patterns. Knowl. Inf. Syst. **6**(3), 243–268 (2004)

RoADS: A Road Pavement Monitoring System for Anomaly Detection Using Smart Phones

Fatjon Seraj[1]([✉]), Berend Jan van der Zwaag[2], Arta Dilo[1,2,3], Tamara Luarasi[3], and Paul Havinga[1]

[1] Pervasive Systems, University of Twente, Enschede, The Netherlands
{f.seraj,a.dilo,p.j.m.havinga}@utwente.nl
[2] Adaptive Systems, Hengelo, The Netherlands
berendjan.vanderzwaag.nl@ieee.org
[3] European University of Tirana, Tirana, Albania
tamara.luarasi@uet.edu.al

Abstract. Monitoring the road pavement is a challenging task. Authorities spend time and finances to monitor the state and quality of the road pavement. This paper investigate road surface monitoring with smartphones equipped with GPS and inertial sensors: accelerometer and gyroscope. In this study we describe the conducted experiments with data from the time domain, frequency domain and wavelet transformation, and a method to reduce the effects of speed, slopes and drifts from sensor signals. A new audiovisual data labeling technique is proposed. Our system named RoADS, implements wavelet decomposition analysis for signal processing of inertial sensor signals and Support Vector Machine (SVM) for anomaly detection and classification. Using these methods we are able to build a real time multi class road anomaly detector. We obtained a consistent accuracy of $\approx 90\%$ on detecting severe anomalies regardless of vehicle type and road location. Local road authorities and communities can benefit from this system to evaluate the state of their road network pavement in real time.

1 Introduction

The technology is riding fast and is spreading everywhere, even in the most remote places where people still face basic road transportation difficulties. The roads will become obsolete with the invention of teleportation, but until then people have to ride on them fast and safe. Meanwhile, the world road network is estimated at $35,433,439$ km[1], and the number of vehicles is estimated at 35 vehicles per 1000 people[2]. Many studies and surveys are made on the topic of roadway deficiencies and their impact on safety and economy [15]. The road surface can wear and deteriorate in time from factors related to location, traffic, weather,

---

[1] CIA World Factbook https://www.cia.gov/library/publications/the-world-factbook/fields/2085.html.

[2] The World Bank 2011 data: http://wdi.worldbank.org/table/3.13.

© Springer International Publishing Switzerland 2016
M. Atzmueller et al. (Eds.): MSM, MUSE, SenseML 2014, LNAI 9546, pp. 128–146, 2016.
DOI: 10.1007/978-3-319-29009-6_7

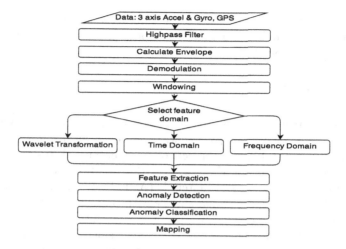

Fig. 1. RoADS flow chart

engineering solutions and materials used to build. In developed countries, Pavement Management Systems (PMS) are specialized structures that handle the duty of the road maintenance. Often these structures are equipped with sophisticated and expensive equipment installed on specialized Pavement Evaluation Vehicles[3]. For example, in the Netherlands since the introduction of PMS in early 1989, 75 % of all of the local authorities, municipalities and provinces, are utilizing a PMS [20]. Developing countries often lack this kind of technology and the $know - how$. They sustain their road network through inefficient financial and maintenance planning. One way to assess the road pavement is to measure a vehicle's vibration with inertial sensors found on smartphones. Therefore, we exploit the pervasive and *'smart'* nature of smartphone devices to collect, process and share these data.

Other research teams are focused more on pothole detection. Potholes are often results of neglected or bad constructed road segments [14]. We are interested to detect and classify more road surface events. Doing so we can monitor in real time the state and the deterioration of the road segments. Figure 1 shows an overview of our solution, from pre-processing, domain selection and feature extraction to anomaly detection and classification.

In this paper we describe RoADS, a smartphone based **R**oad pavement **A**nomaly **D**etection **S**ystem. Section 2 gives a brief overview of the existing works related to road pavement analysis and anomaly detection. Section 3 outlines the analysis and methodology approach to the problem of road anomaly detection. On Sect. 4 we explain our data collection setup, the locations and the methods used to label the collected data, the preprocessing steps, features extracted from different transformations and the method used to reduce the speed and other dependencies from the sensors signal. Section 5 describes the

---

[3] Pavement   Evaluation   Vehicle   https://www.fhwa.dot.gov/research/tfhrc/labs/pavement/index.cfm.

methods and the tools used to classify the data, as well as the obtained results for the labelled and unlabelled roads. Section 6 discusses the conclusion and delineates the future plans.

2   Related Work

To determine the road roughness, road engineers measure the profile of the road. A profile is a segment of road pavement, taken along an imaginary line. Usually the longitudinal profiles are subjects of study because they show the design grade, roughness and texture of the profile [19]. Road roughness is defined by American Society of Testing and Materials (ASTM) [1] as: The deviations of a pavement surface from a true planar surface with characteristic dimensions that affect vehicle dynamics, ride quality, dynamic loads, and drainage, for example, longitudinal profile, transverse profile, and cross slope. Equipment and techniques for roughness estimation are usually categorized into [19]:

- Road and Level survey, surveys performed by a survey crew.
- Dipstick profiler, a hand-held device commonly used for calibration of complex instruments.
- Response type road roughness meters (RTRRMS), transducers that accumulate suspension motions.
- Profiling Devices, sophisticated inertial reference systems with accelerometer and laser sensors to measure the vehicle displacement.

The following works show that inertial sensors alone could be used to detect road surface anomalies. Mainly they were trying to detect potholes, because they are the main concern and also they are relatively easy to detect based on the energy of the event. Studies and practice suggests that potholes are created as a result of distresses and poor drainage of the road surface [14]. Detecting and classifying more of these distresses it will be possible to maintain the long term performance of the road pavements. Beside the anomaly detection, all these works faced one major issue, namely the vehicle velocity. The same road anomaly shows different frequencies and amplitudes when driven over with different speeds. Figure 2 shows the signal generated by a manhole when approached with low and high speeds. An important aspect worth mentioning is also the data labelling method used to annotate the road anomalies. It is crucial to train a precise detection algorithm with the right anomalous segments.

Pothole Patrol (P^2) [5] uses a high resolution 380 Hz accelerometer and a GPS device attached to the vehicle dashboard to collect data and to detect the potholes. Data are transferred to a central server for further processing. Clustering is used to increase detectors precision. Five filters are used, one of them called *z_peak* tries to detect potholes from other high-amplitude road events. Filter *speed vs. z_ratio* is introduced to reject signals with a peak less than a factor t_s times the speed of travel. The labelling technique is based on a trained labeller sitting inside the vehicle and pressing keyboard keys corresponding to predefined anomaly types when they occur.

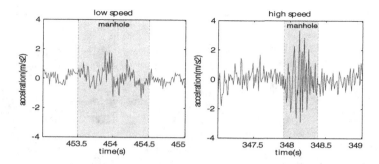

Fig. 2. A 2 s sample of the vertical accelerometer signal when the vehicle drives over a manhole at low speed and at high speed

Nericell [16] uses the Windows Mobile Phone microphone and GPS in conjunction with a high resolution Sparkfun WiTilt accelerometer clocked at 310 Hz to monitor traffic and road pavement. The same technique as in P^2 [5] is used to threshold the acceleration signal and to deal with speed. The novelties consist of introducing another filter named z_sus for speeds <25 km/h, arguing the same anomaly has different shapes for different speeds, and virtual orientation of the phone, using Euler angles to reorient accelerometer data. However, they do not mention the labelling technique they used.

Perttunen et al. [18] use a Nokia N95 mounted on the wind-shield, with accelerometer sampling at 38 Hz and GPS to collect the data. Their algorithm classifies the anomalies into two classes: mild and severe. A method of linear regression is introduced to remove the linear dependency of the speed from the feature vector. Labelling is performed with a camcorder attached to the headrest of the front passenger seat, however they realised this method was unreliable to detect the anomalies. A FFT transformation of the signal is performed to extract frequency domain features and to label the data by plotting together the power spectrum and time domain data. Unclear remains the fact how a 38 Hz accelerometer sensor can generate 17 frequency bands with 1.4 Hz bandwidth.

Tai et al. [21] use a motorcycle riding strictly at two different speeds, 30 km/h and 40 km/h. An HTC Diamond with accelerometer sampling at 25 Hz and a GPS was used to collect the data. Data are preprocessed by the device and sent to a centralised server for classification. Two classification procedures are performed, one to detect the anomalies and the other to rate the road pavement quality from a predefined model of a smooth road. Labelling is performed by the motorcycle rider with a microphone, while riding through an anomaly. An algorithm is used to shift the audio label to correspond with the nearest anomaly event captured by the accelerometer.

3   Methodology

Next we describe the methodology followed during our study.

Transversal   Mild road anomalies   Severe road anomalies

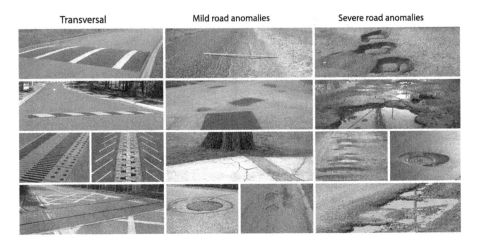

Fig. 3. Types of road anomalies

3.1   Road Pavement Anomaly

The road pavement is characterized by an asphalt or concrete layer surface that facilitates the movement of the wheeled vehicles. Every situation where this pavement disrupts the smooth flow of the vehicle traffic is considered an anomaly. Figure 3 shows different type of road pavement anomalies. Transversal anomalies include: speed humps, speed bumps, road and bridge joints, railroad crossings. As mild anomalies we can classify those segments of road where the surface of the road is not even but potholes are not created yet. They include sunken crocodile cracks, patched roads, tree roots, sunken manholes, raveling (delamination of the sealing layer). Severe anomalies include different types of potholes and severe road deteriorations.

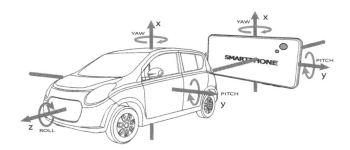

Fig. 4. Smartphone orientation inside the vehicle.

3.2 Vehicle Dynamics

Vehicles are complex dynamic spring-mass systems. Figure 4 shows the orientation of the phone inside the car. All the axis are reoriented with respect to those of the smartphone. It can be observed that in a vehicle exist two forms of movements: the Translational and the Rotational movement.

When analyzing the effect of pothole on the vehicle, only the bounce of the vehicle is taken into consideration. One method to measure the bounce is by measuring the vertical acceleration of the vehicle. However, when a car drives over a road anomaly rotational movements also occurs.

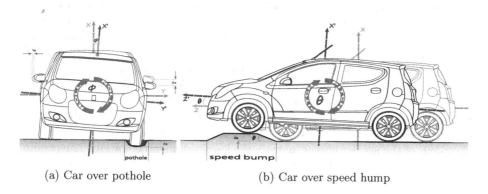

(a) Car over pothole   (b) Car over speed hump

Fig. 5. The dynamics of the vehicle over side and transversal anomalies.

When driving with a constant speed through a smooth flat road, an accelerometer mounted inside the vehicle should only measure the gravity. Once the vehicle hits a pothole as in Fig. 5a the accelerometer should measure the vehicles lateral Y_a and vertical X_a displacement, equal to the depth of the pothole. Also the orientation of the vehicle changes, rolling aside with an angle ϕ. However, if the vehicle drives through a transversal anomaly as in Fig. 5b, the displacement X_a equal to the height of the speed hump, is measured only on the vertical axis X of the accelerometer. Moreover, the orientation of the vehicle changes, pitching up with an angle θ. From these examples we can infer that measuring also the rotational movement of the vehicle can increase the classification rate between the anomalies. MEMS gyroscopes are devices that measure the rotation around a specific axis, the *angular velocity*, and are becoming native to smartphones, like accelerometers and compasses.

3.3 Speed Dependency

The vehicles are equipped with suspension and dampers to attenuate certain vibrations caused by road anomalies. When the vehicle drives slowly, the wheel rpm is low, resulting in less vibration input from the road. Increasing the speed, the wheel rpm and the frequency of the road input also increases, resulting in a vibrating signal as in Fig. 2. We call this phenomenon *speed dependency*. It was noticed that when speed signal from GPS and the absolute values of accelerometer x-axis, for a flat road, are plotted as in Fig. 7a, the envelope of accelerometer follows the speed signal from GPS. Another picture showed up on a non flat road. Figure 6 shows that in the segment between vertical dashed lines the vehicle is driving with high speed through a relatively flat highway, but the peaks are much higher in the segment between vertical black bars, when the vehicle is driving slower up and down a hill. Knowing the vehicle velocity is important to compensate for the speed related variations of the accelerometer. Using a device equipped with GPS, one would assume speed and position will always be available. This is not the case with GPS chips found in smartphones. GPS sensors suffer TTFF (time to first fix) delay. Requirements for a fast TTFF are a clear line of sight and a stable position. To improve the start-up performance of GPS, mobilephones use a technology called A-GPS (Assisted GPS) [11]. A-GPS requires an active data connection with the mobile network operator to receive a preloaded list of available GPS satellites for that location. We did not have an active data connection in our mobile phones at the time we collected the data. On some measurements, data from GPS including the speed were absent. To overcome the absence of speed signal we estimated the velocity by integrating the z-axis of the accelerometer, namely the longitudinal acceleration. The results were satisfactory for the flat Dutch roads, but not for the Albanian roads where we were confronted with slopes. The segment between the black bars in Fig. 6 shows the vehicle climbing a hill, the integrated velocity is increasing but not the GPS speed. The way the speed, slopes and drift behave on inertial sensors readings has a resemblance with the amplitude modulated signals in radio technology [8]. The accelerometer signal measuring complex mechanical vibrations is modulated with different signals, it acts as a carrier for different signals like speed, slope degree, engine and tire revolutions. Envelope demodulation is often used for empirical mode decomposition (EMD) [6] of complex mechanical vibrations in Hilbert-Huang transform [10]. Demodulated signal dS is computed from the raw signal S based on Eq. (1),

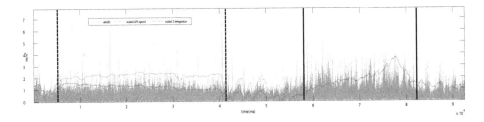

Fig. 6. Readings of linear accelerometer axis $|X|$ (green) with speed from GPS (red) and velocity computed from linear acceleration (blue). Notice the drift accumulated in the segment between two black bars (Color figure online).

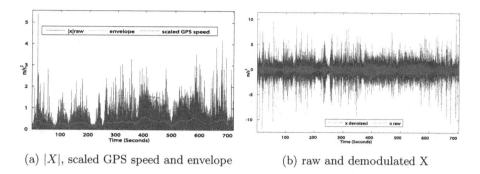

(a) $|X|$, scaled GPS speed and envelope   (b) raw and demodulated X

Fig. 7. (a) Absolute value of x-axis accelerometer $|X_{acc}|$ plotted against scaled GPS speed and envelope, (b) Demodulated signal dX_{acc} (blue) plotted against raw signal X_{acc} (red) (Color figure online).

$$dS(t) = \frac{H \circ S(t)}{E \circ |H| \circ X_{acc}(t)} \tag{1}$$

where t is time, $\circ$ is the function composition, E is a moving average filter, see Eq. (2), with a large window (M = 2000 samples rolling window), H is the high-pass filtered applied to the signal and X_{acc} is the x-axis accelerometer.

$$E(t_i) = \frac{1}{M} \sum_{j=0}^{M-1} X(t_{i+j}) \tag{2}$$

The envelope demodulation does not only represent the speed signal but it also compensates for the slope and other low frequency components. From Fig. 8 is clear that the demodulated signal shows better the anomalies. Figure 7a shows the signal for the absolute values of x-axis, the speed and the envelope. Figure 7b show how the demodulated signal differs from the raw signal.

3.4   Signal Analysis

Once the speed dependency is removed, the amplitudes generated by the anomalies are more uniform. Using a threshold would seem a good approach. However, not all the vehicles generate the same amplitude when driving through the same anomaly. To better understand this process the study can be carried into frequency domain or wavelet transformation, via the signal decomposition into frequency bands.

All the vibrations inside the vehicle happen at a certain frequency bands. The most prominent vibrations on a traveling vehicle are the engine revolutions and the wheel revolutions. A vehicle engine usually operates at a range of 2000 rpm to 3500 rpm or 30 Hz to 50 Hz. When the vehicle is standing still the engine operates at 1000 rpm or 15 Hz. It is common for the city vehicles of B and C segment to have 14 in. to 15 in. tires and to drive in the range of 30 km/h to

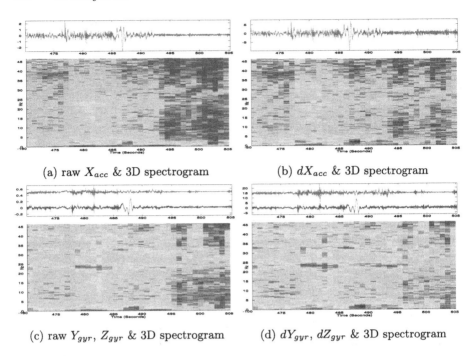

(a) raw X_{acc} & 3D spectrogram   (b) dX_{acc} & 3D spectrogram

(c) raw Y_{gyr}, Z_{gyr} & 3D spectrogram   (d) dY_{gyr}, dZ_{gyr} & 3D spectrogram

Fig. 8. (a,b) 3D spectrogram of raw and demodulated accelerometer signal with plotted x-axis accelerometer (red) on top; (c,d) 3D spectrogram of raw and demodulated gyroscope signal with plotted y-axis (red) and z-axis gyroscope (blue) on top (Color figure online).

90 km/h with wheels spinning at 7 Hz to 23 Hz. The frequency of the vibration from the pothole depends on a lot of factors, the radius and width of the tire, the suspension characteristics, the mass of the vehicle etc.

Road engineers consider road pavement anomalies as longitudinal waves. They all generate vibrations with frequencies depending on the speed of vehicle.

First we tried to use the wavelet transformation as a de-noising tool for the signal. Considering the properties of wavelets and their ability to represent the signal in time and frequency, we decided to use the wavelet transformation for signal analysis. Discrete Wavelet Transform (DWT) uses multi-resolution filter banks and wavelet filters to analyse and synthesise the original signal [4]. It gives frequency resolution in low frequencies and time resolution in high frequencies [12]. Hesami and McManus [9] showed that DWT analysis outperforms power spectral density (PSD) analysis when used to estimate and analyse road roughness.

In the situations when the vehicle is standing, for example at a parking lot, when the vehicle is waiting for a traffic light or on a traffic jam, the vibrations are minimal, mainly the sensor noise. The variance of the low frequency signal is close to zero. Figure 9 shows the variance of the 4-th level of approximation, for vehicles in Trip 1 and Trip 2 respectively, see Table 2. The amplitude of the signal is different. These signal segments can very well be discarded.

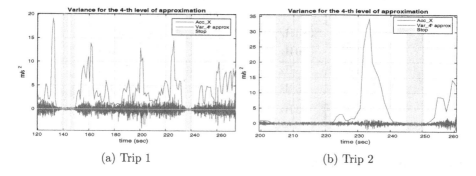

(a) Trip 1   (b) Trip 2

Fig. 9. Stop segments and the variance of the 4-th level of decomposition

3.5 Anomaly Detection

For the anomaly detection supervised machine learning techniques can be applied. The process consists on training algorithms to build a model with a set of training data, and assign the new data to one of the classes. All the data can be represented as distinctive features extracted from the original data. This features can be mapped in space in a way that can be divided into classes. Support Vector Machines (SVM) are algorithms that increase that separable gap as wide as possible considering the margins of the class as support for the gap. SVM assigns weight to the features. This is a suitable approach for the purpose or road anomaly detection, based on the fact that the dynamics of the anomaly are similar for the majority of vehicle, but the amplitudes and the frequency of vibrations vary. The labeling is an important step when training the SVM, it requires the training sets to represent the anomalous signal. Because the anomaly happens for a fraction of time, pinpointing it is rather a difficult process. Audio-visual methods can be used to label the road segments, meticulously annotating every event that happens during the trip. Finding the best parameters during training phase without overt-fitting is crucial, especially in road anomaly detection where the false positive reported potholes cost time and money to the maintenance teams.

4 Data Processing

This section describes our data collection setup, the study areas, the type of anomalies, the technique used to label the data, filtering and feature extraction. Throughout this paper we use the term data referring to the streaming data captured from smartphone inertial sensors.

Table 1. Distance between two consecutive accelerometer measurements at different speeds for different systems.

System	Sampling rate	Speed		
		25 km/h	50 km/h	75 km/h
P^2 [5]	380 Hz	1.8 cm	3.6 cm	5.5 cm
Nericell [16]	310 Hz	2.2 cm	4.5 cm	9.0 cm
Perttunen [18]	38 Hz	18.3 cm	36.5 cm	54.8 cm
Tai [21]	25 Hz	27.7 cm	55.4 cm	83.1 cm
Our setup	93 Hz	7.0 cm	14.0 cm	21.0 cm

4.1 Experimental Setup

Our setup consists of a Samsung Galaxy S2 smartphone running Android 4.0 and an Inertia ProMove 3D[4]. The devices were fixed on the windshield of the vehicle with Nokia Universal Holder CR-114. The Android API does not allow to directly access the sensors used in smartphone, however it allows to choose between five predefined delay intervals at which sensor events are sent to the application: UI, Normal, Game, Fastest or user-defined delay [7]. Therefore two modes were used: Game corresponding to ≈47 Hz, and Fastest corresponding to ≈93 Hz on the Galaxy S2. The Inertia Node was clocked at 200 Hz and was used for testing purpose only.

Data collection software registered independently every available sensor on the phone, with corresponding timestamp in nanoseconds since uptime. GPS timestamp is milliseconds since January 1, 1970 [7]. The drive was recorded in a video using the camera and the microphone of the smartphone. All sensor timestamps were synchronised. Table 1 shows how our setup of 47/93 Hz compares to other setups in capturing small dimension anomalies.

4.2 Collected Data

Data was collected using five different types of vehicles from different roads in two different cities: in and around the city of Vlora in Albania and in and around the city of Enschede in the Netherlands (see Table 3). The selected roads that were used to collect the data represent different types of anomalies. The Dutch roads are in better shape, are flat without many turns. Common anomalies are manholes, speed humps, patches, cracks, bumps and some small potholes. In contrast, the Albanian roads have a lot of slopes, different types and sizes of potholes, bumps and segments of fully deteriorated or unpaved roads. In total we collected data over a distance of 100.3 km on 45.9 unique km of road (see Table 2). Data was collected from accelerometer, gyroscope and GPS sensors of the phone.

---

[4] Inertia ProMove 3D Motion Tracking: http://inertia-technology.com.

Table 2. Total road coverage in km with different vehicles

Trip	Car	Type	Km	Area	Hz	Location
1	Toyota Corolla	hatchback	22.8	rur/urb	47	Enschede(NL)
2	Peugeot 306	hatchback	22.8	rur/urb	96	Enschede(NL)
3	Skoda Fabia	supermini	16.3	urban	47	Vlora(AL)
4	Toyota Yaris	supermini	33	hway/rur	47	Albania
5	BMW X3	suv	5.46	urban	47	Vlora(AL)
	Total		100.3			
	Unique		45.9			

4.3 Data Labelling

We went multiple times over the same road segment, every time with different speed and tried to hit the anomaly in different angles. Labelling was performed by one person using the microphone and the camera of the phone. A detailed list of all known road surface anomalies was compiled. As the vehicle approached the anomaly the labeller mentioned loudly the type of anomaly. The labeller mentioned everything relevant he saw or felt while sitting beside the driver inside the vehicle. We captured the video footage of the trips in 480×720 pixel resolution. The video footage was used to prove the ground truth. Data collected in Albania was not used for training as they were not voice labelled, and labelling them only by video is inaccurate and time consuming.

Urusoft Subtitle Editor[5] is an application used to transcript the movie subtitles. This application was used to transcript all voice labels with accuracy in the order of milliseconds. The subtitle file is synchronised with the smartphone data timestamp, $t_{acc}^n \leq t_{lab} < t_{acc}^{n+1}$ where t_{acc} is the accelerometer timestamp and t_{lab} is the subtitle time, resulting in an array $[t_{lab_{start}}, t_{lab_{stop}}, label]$.

A lag between labelled segments of the data and the actual anomaly was noticed as shown in Fig. 10 where Red signal is shifted from or does not correspond with any peak. Also the video footage was shorter and behind, in time about 1s, than accelerometer data length. Because Android is not a real-time operating system (RTOS), data sometimes may drop when the device is busy [13] and some measured data values may have been delayed, resulting in incorrect timestamps. All lags were corrected manually.

The number of individual anomalies was not sufficient for a successful training. We decided to divide the anomalies into event classes. The anomalous data were divided into 3 event classes.

1. Severe, in this category we labelled sunk-in manholes, small potholes and deteriorated and heavily patched road segments.
2. Mild, in this category we placed all those anomalies that happened only in one side of the vehicle such as cracks, one side patches, one side bumps.

---

[5] http://www.urusoft.net/products.php?cat=sw.

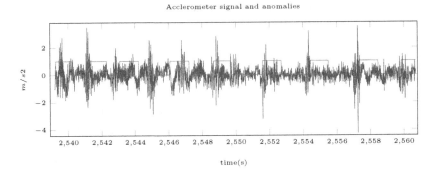

Fig. 10. Signal in blue and binary label (1 for anomaly and 0 for normal road) in red, generated from the voice recordings (Color figure online).

3. Span, in this category were placed road-wide (transversal) bumps, road expansion joins, patches across the road, thick paint, bumps across the road and speed humps.

4.4 Data Filtering

A high-pass filter was applied to the signal to remove the low-frequency components, such as turns, acceleration, deceleration, etc. A first order butterworth highpass filter is designed and applied. Before segmentation, windowing, the speed dependency is removed from the signal using the formula 1.

4.5 Feature Extraction

To prepare data for the classification phase, features representing the data were extracted from accelerometer and gyroscope sensor signals. They are computed from time domain, transformation in frequency domain and wavelet decomposition.

Data was windowed with 256 samples, corresponding to 2.5 s and a 170 sample overlap for the data sampled at 93 Hz, and 128 samples with 85 samples overlap for the data sampled at 43 Hz.

Time-domain features extracted: mean, standard deviation, variance, peak to peak, root mean square, zero crossing rate, mean of absolute value, correlation between all axis, tilt angles, wave form length, signal magnitude area.

Frequency-domain features were extracted after FFT transformation with a Hamming window function: mean frequency, median frequency, energy of the frequency bands.

Wavelet decomposition: To decompose the signal we used Stationary Wavelet Transform(SWT) [17] a form of nondecimated DWT. Several experiments were carried out with different wavelet families. Sym5 wavelet from symlet wavelet

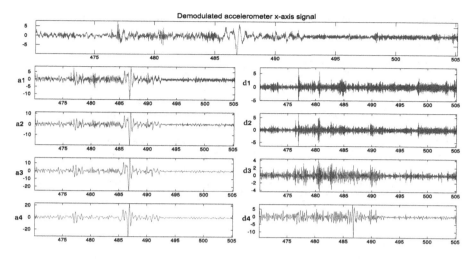

Fig. 11. SWT decomposition at 4 levels with a sym5 wavelet.

family, and 4 levels of decomposition gave the best results. Figure 11 shows the plot of the original signal on top and 4 levels of SWT decomposition, the approximations on the left and the details on the right. The high frequencies are visible at the details column d1 and lower frequency signal at a4. Also because we are using SWT we can a have a good time resolution for every level of decomposition. The following features were extracted from SWT decomposition: absolute mean, standard deviation, variance, energy for every level of detail and approximation. Four levels of wavelet decomposition were appropriate for our signal.

5 Classification and Results

Assuming the number of anomalous windows is lower than the number of normal windows, we used a 2 step classification. As the first step, all the windows are processed to detect the anomalous windows (those containing road events) from the normal ones. In the second step, the anomalous windows are processed through another classifier to classify the type of anomaly.

5.1 Training the SVM

To train the detector we used our labeled data consisting of 3066 windows, 2073 normal windows and 993 anomalous windows. We use a sliding window with an overlap of 66 % to frame the data, which means that parts of the anomalous signal would be present in more than one window. Training was performed using 10-fold cross-validation, where 1/10 of both anomalies and normal data were used only for testing purpose. The training data were not stratified. From 993 anomalous windows, we used only those windows with the anomalous signal in the center of the window spreading equally in both sides. The reason behind

Table 3. The confusion matrix and accuracy for the classification of anomalous (Positive) and normal (Negative) segments of road, with different feature sets.

Confusion matrix anomaly vs Normal												
Method	Accuracy	TP	TN	FP	FN	Spec.	Sens.	Prec.	G	RS	FPR	FNR
TD_{raw}	81.43 %	49	201	39	18	0.82	0.73	0.196	0.77	0.89	0.18	0.26
$TD_{demodulated}$	85.26 %	52	214	31	15	0.87	0.82	0.1955	0.84	0.94	0.13	0.18
FFT_{raw}	77.88 %	49	194	51	18	0.79	0.73	0.2016	0.76	0.92	0.21	0.27
$FFT_{demodulated}$	79.17 %	40	207	38	27	0.84	0.59	0.1619	0.71	0.70	0.15	0.40
SWT_{raw}	82.69 %	50	208	37	17	0.85	0.75	0.1938	0.80	0.88	0.15	0.25
$SWT_{demodulated}$	88.14 %	57	218	27	10	0.89	0.85	0.2073	0.87	0.96	0.11	0.15
$TD+FFT_{demodulated}$	83.01 %	48	211	34	19	0.86	0.72	0.1853	0.79	0.83	0.14	0.28
$TD+SWT_{Demodulated}$	88.78 %	59	218	27	8	0.89	0.88	0.213	0.89	0.99	0.11	0.12

*TD = Time Domain, *FD = Frequency Domain, *SWT = Stationary Wavelet Transform
TP/FP = True/False positive, TN/FN = True/False negative, Spec./Sens. = Specificity/Sensitivity, Prec.. = Precission
G = G-mean, RS = Rlative sensitivity FPR/FNR = False positive/negative rate

this decision, is that using a sliding window method, parts of the same anomaly are present in different consecutive overlapping windows and the aim is to train the detector with the best representative of the class. Smartphones are subject of processing power and energy concerns. We try to keep the number of calculated features as small as possible and in the mean time to be flexible with the number of classes. We already discussed the complexity of the vehicle vibration response on the road, therefore is quite impossible to separate the anomalies by training thresholds. We decided to use a radial basis function (RBF) kernel, as described by Burges [3] and Ben-Hur et al. [2]. Using a RBF kernel the feature set is transformed into higher dimension feature space, where the features are easily classifiable. A grid search was conducted to find the best values for *hyperparameters* [2]: the kernel parameter $\gamma = 0.002$ and 0.0002, the cost of misclassification the soft margin constant $C = 320$ and 100 respectively for the detection and classification. Several experiments were performed, training the detector with feature sets from different domains, transformations and combinations thereof, see Table 3. We also experimented with different settings, such as features extracted from the raw signal TD_{raw}, on features from the demodulated signal $TD_{demodulated}$. The best results were achieved by the TD+SWT, the combined feature set from time domain and wavelet transformation. Nevertheless for detection we used only features from SWT.

5.2   Training Results

The results from the training experiments clearly show that the detectors trained with the demodulated features were more accurate on classifying the data than the detectors trained with raw data. Table 3 shows the results of the anomaly detection.

Table 4 shows anomaly classification results of our 10-fold crossvalidation training of the data with features extracted from the wavelet decomposition (SWT) and combined (TD+SWT) features from time domain and wavelet decomposition. The accuracy is the same 91 %. However, the severe class is

Table 4. The confusion matrix and accuracy of the classification for anomalous segments of road with features from SWT and TD+SWT.

| | Confusion matrix | | | | | |
| | SWT | | | TD+SWT | | |
Class	Sev.	Mild	Span	Sev.	Mild	Span
Severe	17	3	0	16	4	0
Mild	1	15	0	0	16	0
Span	0	0	9	0	0	9
Accuracy	91.1 %			91.1 %		

detected better with features from the wavelet transformation only. We also experimented with features from other transformations but the results were lower than 86 %.

5.3   Evaluation on Unlabeled Data

The evaluation process turned out to be easy using our labeling technique. The system made the predictions and two subtitle files were generated containing the labels of the anomalies detected. The algorithm also generates a KML[6] file with the location of severe anomalies for Google Maps. We went through the video footage taking notes on reported anomalies from the subtitles. From data collected in Albania some types of severe anomalies were not detected. We believe the reason for that is the fact that our system had not been trained for those anomalies. Table 5 shows the performance of our system with data collected with different vehicles in different locations. Trip 1 and Trip 2 are made in the Netherlands, in the same road used for training the system. Trip 3, 4 and 5 are trips made in Albania (see Table 2). The trips in the Netherlands have a higher ratio of anomalies per total number of windows than those in Albania. The reason is the driver behaviour in our tests. The video footage showed that the drivers in Albania tried to avoid damage by swerving around or slowing down in front of bad road areas, whereas the drivers in the Netherlands did not. The GPS location from mobile phone on the map is not very accurate (see Fig. 12). In a wider perspective with more data from different contributors it create the possibility for clustering all measurements by their geo-coordinates.

To estimate the detector performance for the severe class we counted the false negative windows reported by the detector and also the undetected anomalies from the video footage, see Table 5. The detector detects all windows with anomalous signal, including consecutive windows for the same anomaly. To avoid clutter on the map, if two or more consecutive windows detected the same anomaly, we keep the half of them. This way the number of anomalies shown on the

---

[6] Keyhole Markup Language https://developers.google.com/kml/documentation/ kmlreference.

Table 5. The results of system classifications with models and features from Time Domain and Wavelet Decomposition

				Detection	Classification			Mapping	Evaluation			
		S/R	#Win	Anom	Sev.	Mild	Span	mapped	Sev.	TD	FD	MA
Trip 1	TD	47 Hz	609	195	107	68	20					
	SWT	47 Hz	609	165	119	26	20	113	73	64	15	4
Trip 2	TD	96 Hz	764	222	76	105	41					
	SWT	96 Hz	764	207	93	72	42	123	61	52	12	3
Trip 3	TD	47 Hz	1067	280	119	106	55					
	SWT	47 Hz	1067	222	120	51	512	152	82	74	10	8
Trip 4	TD	47 Hz	2240	584	349	176	61					
	SWT	47 Hz	2240	412	255	98	59	296	173	no video		
Trip 5	TD	47 Hz	794	239	113	126	0					
	SWT	47 Hz	794	192	99	93	0	134	76	74	6	6

Sev. = Severe, TD = True Detections, FD = False Detections, MA = Missed Anomalies
mapped = Nr. of anomaly points pinpointed on the map

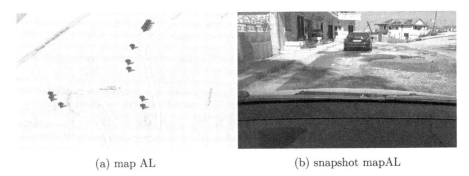

(a) map AL   (b) snapshot mapAL

Fig. 12. Snapshots of map and video anomalies for locations in Vlora, Albania.

map is fewer than the number of reported windows. As false detections (FD) were counted all severe anomalies that were detected as mild ones, but that can be subjective based only on video footage. Missed anomalies (MA) is the number of all anomalies, counted by us through the video but not reported from the detector. For Trip 4 associated with Fig. 6 the video file was corrupted, but based on GPS map we noticed that the majority of the anomalies were in the segments of deteriorated road, the segment between black lines. For Trip 5, made with the BMW X3 SUV, the results were more correct. Worth mentioning is the fact that the system has a high accuracy on detecting transversal anomalies, spanning the width of the road. On extremely deteriorated road segments, the system detected windows that belong to the transversal anomaly class, see Table 5. In comparison, P^2 [5] and Nericell [16] skipped the windows representing transversal anomalies.

6   Conclusions

In this paper we proposed a system that detects road surface anomalies using mobile phones equipped with inertial sensors: accelerometers and gyroscopes. We applied the stationary wavelet transform analysis and a method to remove effects of speed, slopes and drifts from sensor signals using the envelope technique. Our audiovisual labeling technique was precise and also helpful for the system evaluation. Classifying road anomalies is a rather difficult process and the expectancies to detect all road anomalies on one pass are quite low. Nevertheless, the obtained results showed a consistent accuracy of ≈90 % on detecting severe anomalies regardless of vehicle type and road location. To increase the accuracy and the number of anomaly classes we will collect more labeled data and improve the training of the SVM through stratification and testing with other feature sets and sensors, such as barometric pressure sensor. For future work, we aim to apply these methods of road anomalies detection in participatory sensing using clustering by the geo-coordinates. We will look and try to address the time consuming process of manual labeling, by automatic recording and labeling bigger datasets. To address the problem when the driver steers away from the anomaly a driver behavior detection should be investigated. Also in cases when the drivers intentionally skip the anomaly, driver behavior detection algorithms can be implemented. We also intend to implement a vehicular network to share that information with other vehicles and to perform road serviceability performance with outputs conform the International Roughness Index and the ISO 2631 standard.

References

1. ASTM Standard E867, Standard Terminology Relating to Vehicle-Pavement Systems, June 2012
2. Ben-Hur, A., Weston, J.: A users guide to support vector machines. In: Carugo, O., Eisenhaber, F. (eds.) Data Mining Techniques for the Life Sciences, Methods in Molecular Biology, vol. 609, pp. 223–239. Humana Press, New York (2010)
3. Burges, C.J.: A tutorial on support vector machines for pattern recognition. Data Min. Knowl. Disc. 2(2), 121–167 (1998)
4. Daubechies, I.: The wavelet transform, time-frequency localization and signal analysis. IEEE Trans. Inf. Theory 36(5), 961–1005 (1990)
5. Eriksson, J., Girod, L., Hull, B., Newton, R., Madden, S., Balakrishnan, H.: The pothole patrol: using a mobile sensor network for road surface monitoring. In: Proceedings of the 6th International Conference on Mobile Systems, Applications, and Services, MobiSys 2008, pp. 29–39. ACM, New York (2008)
6. Feldman, M.: Signal Demodulation. Wiley, New York (2011)
7. Google: Android developer sensor and location classes. http://developer.android. com/reference/android/hardware/Sensor.html
8. Gottlieb, I.: Understanding amplitude modulation. Foulsham-Sams techn. books, H. W. Sams (1966)
9. Hesami, R., McManus, K.J.: Signal processing approach to road roughness analysis and measurement. In: TENCON 2009–2009 IEEE Region 10 Conference, pp. 1–6. IEEE (2009)

10. Huang, N., Attoh-Okine, N.: The Hilbert-Huang Transform in Engineering. Taylor & Francis, New York (2005)
11. LaMance, J., DeSalas, J., Jarvinen, J.: Innovation: assisted GPS: a low-infrastructure approach. GPSWorld **13**, 46–51 (2002)
12. Mallat, S.: A Wavelet Tour of Signal Processing: The Sparse Way. Academic Press, New York (2008)
13. Milette, G., Stroud, A.: Professional Android Sensor Programming. Wrox, Birmingham (2012)
14. Miller, J.S., Bellinger, W.Y.: Distress identification manual for the long-term pavement performance program (fourth revised edition). Technical report FHWA-RD-03-031, Federal Highway Administration, June 2003
15. Miller, T., Zaloshnja, E.: On a crash course: The dangers and health costs of deficient roadways (2009)
16. Mohan, P., Padmanabhan, V.N., Ramjee, R.: Nericell: rich monitoring of road and traffic conditions using mobile smartphones. In: Proceedings of the 6th ACM Conference on Embedded Network Sensor Systems, SenSys 2008, pp. 323–336. ACM, New York (2008)
17. Nason, G.P., Silverman, B.W.: The stationary wavelet transform and some statistical applications. In: Antoniadis, A., Oppenheim, G. (eds.) Wavelets and Statistics, pp. 281–299. Springer, New York (1995)
18. Perttunen, M., et al.: Distributed road surface condition monitoring using mobile phones. In: Hsu, C.-H., Yang, L.T., Ma, J., Zhu, C. (eds.) UIC 2011. LNCS, vol. 6905, pp. 64–78. Springer, Heidelberg (2011)
19. Sayers, M., Karamihas, S.: The Little Book of Profiling: Basic Information about Measuring and Interpreting Road Profiles. University of Michigan. Transportation Research Institute, UMTRI (1996)
20. Schut, P., de Bree, T., Fuchs, G.: Responsible pavement management. In: First European Pavement Management System: Conference-Proceedings and Final Program (2000)
21. Tai, Y.C., Chan, C.W., Hsu, J.Y.J.: Automatic road anomaly detection using smart mobile device. In: Proceedings of the 2010 Conference on Technologies and Applications of Artificial Intelligence (TAAI 2010), 18–20 November 2010, Hsinchu, Taiwan, pp. 1–8 (2010)

Mining Ticketing Logs for Usage Characterization with Nonnegative Matrix Factorization

Mickaël Poussevin, Emeric Tonnelier[✉], Nicolas Baskiotis, Vincent Guigue, and Patrick Gallinari

Laboratoire d'Informatique de Paris 6, UMR 7606, CNRS, Sorbonne-Universités, UPMC-Paris 6, Paris, France
emeric.tonnelier@lip6.fr

Abstract. Understanding urban mobility is a fundamental question for institutional organizations (transport authorities, city halls) and it involves many different fields like social sciences, urbanism or geography. With the increasing number of probes tracking human locations, like RFID pass for urban transportation, road sensors, CCTV systems or cell phones, mobility data are exponentially growing. Mining the activity logs in order to model and characterize efficiently our mobility patterns is a challenging task involving large scale noisy datasets.

In this article, we present a robust approach to characterize activity patterns from the activity logs of a urban transportation network. Our study focuses on the Paris subway network. Our dataset includes more than 80 millions travels made by 600 k users. The proposed approach is based on a multi-scale representation of the user activities, extracted by a nonnegative matrix factorization algorithm (NMF). NMF is used to learn dictionaries of usages that can be exploited in order to characterize user mobility and station patterns. The relevance of the extracted dictionaries is then assessed by using them to cluster users and stations. This analysis shows that public transportation usage patterns are tightly linked to sociological patterns. We compare our approach with a k-means baseline that does not take into account user information and demonstrate the interest of characterizing user profiles to obtain better representations of stations.

1 Introduction

The literature on urban mobility is vast and diverse but until recently, it has mainly focused on explanatory statistics of global behaviors. With the development of tracking techniques such as mobile phone networks, the last decade has seen a multiplication of quantitative statistical studies. For example, frequency scales of travels are analyzed in [3] and multiple studies have shown that it is possible to predict most daily travels [19]. For public transportations, some early studies focused on opinion pool to analyze the modification of user behaviors after the creation of new lines [6]. In this domain, quantitative data are recent

© Springer International Publishing Switzerland 2016
M. Atzmueller et al. (Eds.): MSM, MUSE, SenseML 2014, LNAI 9546, pp. 147–164, 2016.
DOI: 10.1007/978-3-319-29009-6_8

and linked to the adoption of smart cards to authenticate users in most cities, like e.g. in London, Lisbon or Paris. Up to now, they have been exploited for problems like bottleneck detection [4] or frequent pattern prediction [5], which turns out to determine the time and location of the next trip for a given user.

However, no study has focused on mining the temporal and spatial profiles of individual users. Analysis have been performed on Shanghai taxis [16] and Parisian public bicycle sharing system [17] for example, but they have been mainly focused on extracting global statistics and none of them has analyzed individual user traces across multiple travels during multiple days. Discovering latent usage of public transportations is a crucial need for institutional authorities: lot of efforts are spent on conducting ground surveys to achieve a partial understanding of the habits of their clients. Such knowledge is important for pricing policies, load management and planning. We propose to exploit ticketing data to extract regularities in usage patterns in order to identify the hidden activities that causes each individual event (i.e. to explain all logs in the data).

Our analysis focuses on data provided by the STIF (*Syndicat des Transports en le de France*, Paris area transport authority) and contains around 80 millions log entries overs 91 days, with an explicit identifier for both the station (300 locations) and the user (600 k ids). The noise inherent to the individual activity traces and the size of the logs explain why ticketing data has hardly been used. We propose here an experimental study to show the potential of machine learning to exploit transport data. We introduce a user centered multi-scale representation of the data and we use a constrained nonnegative matrix factorization (NMF) to extract latent activity patterns. Based on this extracted representation, we build station profiles that we cluster and analyze the obtained segmentation. We demonstrate the interest of our approach with respect to a k-means baseline: we show that an efficient profiling requires a user modeling step, even if finally we focus on station representation. We also focus on data reconstruction and abnormality detection: our model is able to output log predictions. Using a symmetrised Kullback-Leibler divergence between ground truth and predictions reveals some abnormalities in the log flow associated to each station.

The paper is organized as follows. We briefly review literature on related work in Sect. 2. To face the challenge of data size and sparsity, we propose in Sect. 3 a model to aggregate ticketing logs per user and station on three frequency and two temporal scales. In Sect. 4 we present the nonnegative matrix factorization model used to extract the usages and latent activities from user events. Analysis of the results on our dataset is presented in Sect. 5. As an application case, in Sect. 6 we illustrate how to use this model for clustering stations and extracting correlations between temporal habit and sociological realities.

2   Related Work

We present below a synthesis on the related work both on urban mobility understanding and on nonnegative matrix factorization algorithms.

2.1 Urban Mobility

The problem of understating urban mobility has been studied at different levels. [2] studied city planning policies in order to promote the use of performance indicators for sustainable public transports. Many studies [3,8] have focused on private vehicle travels, using mobile phone networks to track a population and to characterize the time scales of these travels. In [19], the authors showed that most private vehicle travels are predictable and in [22] they also linked travel behaviors and social network behaviors. The recent study [15] also uses mobile phone networks to extract hot spots from week day travels in the 31 biggest Spanish cities while [10] focused on car traffic analysis and abnormality detection. [14] used 6 months of GPS data coming from the 33000 taxis in Beijing, covering an impressive 800 million kilometers, to analyze the causes of possible abnormalities. Also with GPS data but on 2000 Shanghai taxis, [16] used nonnegative matrix factorization to characterize the behavior of taxi drivers. Finally, several recent works have focused on new ways to track users. For example [13] describe the use of Bluetooth scanners to track pedestrians visiting Duisburg zoo. Similarly in Paris, [17] mined spatio-temporal clusters using Paris Velib' data (city shared bicycle service), but without having access to the user identification, they could not track individual users.

The creation of smart cards to authenticate users [6] allows for wider scale and finer analysis. One important target has been the identification of bottlenecks in the network. For instance [4] studied spatio-temporal distribution of users in the subway of London. In the same way, [5] focused on itinerary prediction, during week days for buses, so as to inform users in case of problems in the network, mining a dataset of 24 million travels of 800000 users over 61 days.

In contrast with previous work focused retrieving global traffic information, the analysis presented here focuses on discovering a collection of trip habits that can be used to describe individual users. As in [16], our model relies on a modified nonnegative matrix factorization, described in Sect. 4, to extract behavioral atoms. We also exploit the user identification to characterize travels through the notion of periodicity.

2.2 Nonnegative Matrix Factorization (NMF)

Matrix factorization approaches have long been used in data mining and are well described in [7]. Nonnegative matrix factorizations [1] extract constructive representations over a set of extracted basic components out of nonnegative data. Each basic component is named *atom* and the extracted component collection is named *dictionary*. The dictionary is learned such that each data can be approximate by the positive weighted combination of atoms of the dictionary. A such weighted vector over the atoms is named *code*. Matrix factorization has shown interesting performances as a feature extractor when the nonnegativity is a sensible part of the data either to obtain a composition percentage over a dictionary like in face detection [24] or when a negative coding does not make sense like in topic extractions from document application [18].

Matrix factorizations solve two problems at once: learning the dictionary and the associated code. Constraints are generally added during the learning algorithm as regularization terms. Nonnegativity [12] and sparseness [11] are two common constraints. As no subtraction is allowed with the nonnegativity constraint, Nonnegative Matrix Factorization (NMF) is more likely to obtain part-based representation with atoms being parts of the initial signals and the code being the proportion of each atom in the signal. Sparseness forces the model to reconstruct initial data using only a few atoms.

While a completely different field, our task might be linked to the separation of musical sources [21] or note identification in a music track [20]. The log of a user corresponds to frequent and multiple activities, it can be seen as a stream of pulses generated by multiple sources [9]. However the variability of the time of authentication, which is our reference event, hinders the approaches based frequency representation used in most signal applications. In [23], nonnegative matrix factorization has been adapted to time event detection to identify *time-shift* invariant patterns. It cannot be used as such to detect temporal behaviors as they are typically time dependent: our idea is to characterize activities based on event occurrence.

3   Subway Data Analysis and Modeling

In this study, we aim at discovering patterns in the subway usage of any user in order to characterize each log by a corresponding activity: an event which occurs every working day around 7 a.m. may be categorized as going to work, an event which occurs some Friday night as nightlife, etc. Tagging authentication events with a social activity allows us to characterize both the user (workplace, home, friends' home, recreational habits) and the station at the same time. We propose a model extracting meaningful activity temporal patterns and allowing a categorization of the subway traffic according to different usages. In this section, we first describe the data and then our modeling.

3.1   Ticketing Logs

We use a dataset collected by the *Syndicat des Transports en Île-de-France* (STIF), the transport organization authority in charge of Paris public transport. More than seven millions of Paris transport users have subscribed to a pass managed by the STIF. The ticketing logs record every authentication of any pass with its location and precise time. This dataset provides an accurate real-time picture of the use of a public transportation system. The analysis is challenging for two reasons: the size of the data (5 GB/month) coupled to the sparsity of individual user data hinders the mining of frequent behavioral patterns; secondly, the dataset contains only a subset of the urban network activity. The STIF estimates that 20 % to 30 % of logs are missing, either due to the malfunctioning of a turnstile or to the user voluntary not authenticating itself.

Moreover, the logs correspond to a check-in action, as pass checkpoints are positioned at entry points in metro stations and buses, and not at exit points. Thus the user's itineraries are not explicitly present in the dataset, only a (large) part of their check-ins are recorded.

3.2 Modeling

In the following, u will refer to a user, s to a station and t to a time. An authentication at time t is thus a triplet $\ell = (u, s, t)$. The mobility of a user is partially described by the log of its authentications, $L = \{\ell = (u, s, t)\}$. Each authentication is related to a latent activity of the user. We propose in this section a latent model using a multi-scale aggregation of events by day and by week.

A first difficulty for identifying latent activities lies in the wide frequency scale of the events. Non-frequent events are hidden by frequent ones, any approximation will catch frequent day-to-day activities and miss the less frequent patterns. We propose to dispatch user activities in three frequency bands: high frequency for events occurring more than twice a week, medium frequency for events occurring at least once each 10 days and less than twice a week, and low frequency for unusual events. We use the spatial information to filter events in the frequency bands: for a given user, each event is associated to a frequency band based on the number of times the event station appears in the user log.

We make the assumption that within a frequency band activities are more characterized by their occurrence time than by their location. On the one hand, this assumption correctly interprets regular activities occurring at the same place (like leaving for work, going back home); on the other hand, it allows us to capture recreational activities that occur in wide areas (like going out to restaurant, visiting friends).

Within a frequency band, an event is represented by the couple (u, t). The user behavior can be modeled by the probability function of the authentication time for this user in the three frequency bands: $p^b(t|u)$, $b \in \{low, medium, high\}$. Our goal is to discover a set of activities $\mathcal{A}$ over which the user log can be decomposed as: $p^b(t|u) = \sum_{a \in \mathcal{A}} p(t|a) * p(a|u)$. Our approach relies on a multi-scale representation of authentication events, by day and by week. We chose to characterize an activity, that we denote a, by the probability function of the ticketing event during the day $f_{d,a}(t) = p(t|a)$ and during the week $f_{w,a}(t) = p(t|a)$ with t respectively a day time and a week time variable. As we are more interested on the discovery of widespread usages, our objective is to infer a small set of activities $\mathcal{A}$ sensible for all users.

This formalism supposes a weekly pattern and does not allow to model explicitly the localization information. However, the station information can still be decoded from the individual user data knowing the activity decomposition for a user.

3.3   Notations and Data Representation

We first filter authentication triplets (u, s, t) in three frequencies bands: *low* for couples (u, s) occurring only a few times, *high* for frequent couples (u, s) and *medium* for everything in between. On Fig. 1 is represented the authentication set of a single user with stations sorted by frequency, from low to high. The two most frequent stations for this user are likely to be its home and workplace. For all frequency bands f in $\{low, medium, high\}$, we build a multi-scale vector representation of a user as the concatenation of the two probability functions of the authentication time during the day and the week. Both are empirically estimated using a time step of 15 min for the day and 2 h for the week. Thus each user is represented by a n dimensional vector with $60/15 * 24 = 96$ dimensions for the day and $24/2 * 7 = 84$ for the week. We denote m the number of users. As a result, data are represented by a set of 3 matrices: $\{X^{(b)} \in \mathbb{R}^{m \times n} | \forall b \in \{low, medium, high\}\}$.

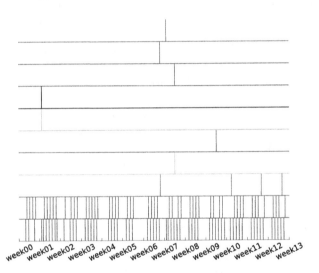

Fig. 1. A user of the Parisian network authenticated at 10 stations over 91 days. Stations are ordered by decreasing frequency from bottom to top. The most frequent stations are relative to the residence and the workplace. Less frequent ones are likely to correspond to recreational activities.

Figure 2 shows a subset of user profiles aggregated without the frequency filtering. The data is noisy reflecting users' individual variance. Still this representation extracts peaks of activities. As argued above, without the frequency filtering most of the density is used by recurrent commuting patterns. User profiles after frequency filtering are represented in Fig. 3, where *low*, *medium* and *high* are respectively the bottom, middle and top. Typically, commuting patterns are present in the high frequency band and week-end and evening events are present in the other two bands. Some users have no event in a particular band, which is a strong characterization of their behaviors.

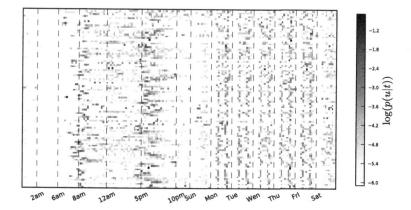

Fig. 2. Aggregated user profiles over time ordered by time of their highest peak. Concatenation of the day and week scales.

4 Learning Usage Atoms with NMF

The previously extracted matrices from authentication data represent the averaged daily and weekly user profiles. We aim at extracting the latent behavioral patterns which compose these profiles. Our goal is to achieve a granular identification of behavioral patterns, allowing us to characterize heavy and light traffic periods, during evening, nights and week-end. Thus the challenge here is to extract generic patterns reflecting the most common patterns like commuting but also less frequent ones happening during evenings and week-ends without over-fitting and learning patterns specific to only of small set of users.

We propose to use a nonnegative matrix factorization algorithm to extract the pattern dictionary with a mono-modal constraint on dictionary rows. As nonnegative matrix factorization decomposes each sample as a positive linear combination of atoms, it is well-suited to our problem: we seek to approximate a user by a set of complementary behaviors.

Formally, the goal is to approximate each matrix X by the product of two positive matrices, D the pattern dictionary and A the code matrix: a row of D, named atom, represents the time profile $p(t|a)$ of a particular activity a and a row of A contains the coordinates of a user in this usage space, that is the repartition $p(a|u)$, for the user u, of the activities in $\mathcal{A}$. These coordinates can be viewed as the usage probabilities for the user.

We want to take into account the following constraints:

– normalization: each dictionary atom is the concatenation of the daily and the weekly information representing the probability density function of the ticketing event; thus both parts of the dictionary atom has to sum up to 1;
– mono-modal atoms: an atom is supposed to explain a unique activity in the day, localized at a precise time window;
– sparsity of the reconstruction: a user has naturally few activities, thus only a small set of dictionary atoms should have a non-zero weight in each row of A.

The dictionaries for the three frequency bands are distinct: they can be learned separately. For each frequency band, the problem can be formalized as the minimization of a loss function $\mathcal{L}$ under the constraint $C(D)$ on the dictionary D:

$$\mathcal{L}(X, A, D) = \frac{1}{m}\|X - A.D\|^2 + \lambda|A| \tag{1}$$

$$C(D) : D \geq 0, \forall i, \sum_{j<t_{day}} D_{ij} = 1, \sum_{t_{day}\leq j<t_{week}} D_{ij} = 1 \tag{2}$$

The NMF optimization is done using a projected gradient (with projection ϕ on the constraints) for the dictionary and multiplicative update rules for the codes (see Algorithm 1).

$D, A \leftarrow$ rand ;
while *not converged* **do**
    $D = D - \mu(A^T(X - AD));$
    $D = \phi(D)$;
    $A = A \odot \frac{XD^T}{\lambda+ADD^T}$;
end

Algorithm 1. NMF update rules: projected gradient on D and multiplicative update rules on A

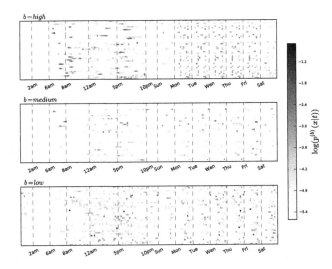

Fig. 3. Aggregated user profiles overtime with frequency filtering. Events with high, medium and low frequencies are in top, middle and bottom charts respectively. Concatenation of the day and week scales.

To force the mono-modal property of dictionary atoms, an additional projection is used every 100 iterations, by applying a Gaussian filter to each atom in order to capture only the highest peak of the daily part:

$$\forall i, \quad d_{i,day} \leftarrow d_{i,day} \odot \exp(-\frac{(t_{day} - t_{peak})^2}{2\sigma^2}) \tag{3}$$

where $\odot$ is the element-wise multiplication, t_{day} ranges over the day dimensions of the atom, t_{peak} is the time corresponding to the peak day time in the atom and σ a fixed parameter defining the granularity of the time window and $d_{i,day}$ is the part of i-th row of the dictionary D accounting for daily patterns.

5   Analysis of Extracted Representation

We focus our analysis on the subway traffic, excluding buses, trains and trams. As we are looking for mobility usages. Users were filtered to retain only those with enough travels to have a significant activity and subscribing a monthly pass. Our dataset is composed of around 80 million user authentications at one of the 300 stations over 91 days for the subway network of the city of Paris, concerning around 600 k unique users.

The $k = 100$ atoms were extracted with 180 features of $m = 600000$ users, using a 8 cores (3.07 GHz) and 16 GB of RAM PC. It takes approximately 10 h to complete the 1 k iterations of the nonnegative matrix factorization algorithm.

Extracted atoms are represented in Fig. 4 where they are sorted by their occurrence of highest peak. In the high frequency band, where commuting patterns are the most frequent ones, the mono-modality allows the discovery of joint *leaving-for-work* and *going-back-home* patterns. The extraction is mainly focused on working days since only a small part of the density occurs during week-ends. With most atoms occurring in the morning the high frequency band has a fine granularity on the *leaving-for-work* usage. In the medium frequency band more atoms are devoted to lunch activities and they typically are active during working days too. Still some atoms, bottom ones, are catching evening and week-end activities. In the low frequency band, density of the week is more evenly distributed through the seven days. A vast majority of the atoms are representing evening and nightly activities. Note that the working hours of the subway are 5:30 am to 2 am, with a slight variance over days, which explains why there is no activity corresponding to the 2 am to 5:30 am period.

Quality measures of the nonnegative matrix factorizations are presented in Table 1. The data sparseness is the average percentage of nonzero entries per user. The dictionary sparseness is the average percentage of nonzero entries per dictionary atom. The code sparseness is the average percentage of nonzero entries per user code (a row of α). The important activities row counts the average number of activities responsible of 90 % of a user's profile. It is interesting to see that the factorization on the *low* frequency band scatters the density over more activities than the *medium* and *high* band. It is coherent with the fact that infrequent validations correspond to more diverse usages (visiting friends, hiking

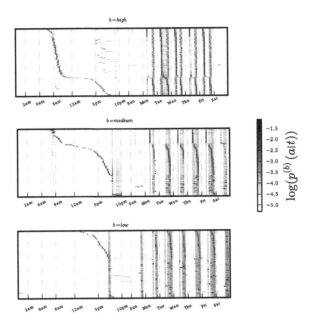

Fig. 4. Activity atoms extracted from the profiles per frequency band. Concatenation of day and week scales. Each row is an atom, sorted by the hour of highest peak.

Table 1. NMF sparsity measures on the *low*, *medium* and *high* frequency bands

NMF	*low*	*medium*	*high*
Data sparsity (std)	16.98 (0.38)	7.97 (0.27)	16.88 (0.37)
Dictionary sparsity (std)	51.15 (0.50)	52.41 (0.50)	51.93 (0.50)
Code sparsity (std)	9.23 (0.29)	4.33 (0.20)	5.46 (0.23)
Important activities (std)	4.38 (0.20)	1.94 (0.14)	1.68 (0.13)

in the city, going to a restaurant,...). This value has to be put in perspective with the bar plots and histograms of Fig. 5. The bar plots of the top row of Fig. 5 represents, per frequency band, the percentage of users that have a non-zero weight attributed to each dictionary atom. The shape of the curves indicates that all dictionary atoms are evenly used to describe users. This is the sign that the model did not over-fit the dataset and it is one of the property we were looking for: the extracted behavior pattern, the dictionary atoms, are not to specific to one particular user. The bottom row histograms count the number of dictionary atoms per user. The first noticeable effect of the frequency filter is that in the medium and high frequency bands a great proportion of users are characterized by a lack of events. It also confirms the good sparsity of the extracted representation: the NMF uses few dictionary atoms to reconstruct each user profile. NMF appears as a robust solution to extract latent activity patterns from noisy temporal data.

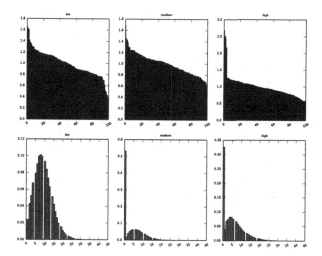

Fig. 5. Top row represents usage of each dictionary atom over the set of users per frequency band. Bottom row contains the histogram of the number of dictionary atoms having non-zero weights per user per frequency band.

6   From Users to Stations

Using NMF, each user is now represented as a vector of usage weights. In this section we analyze the distribution of the individual usage patterns according to the metro station geographical position. Formally, we want to estimate the probability function $p(a|s)$ of activities over subway stations. We will use the following decomposition: $p(a|s) = \sum_u p(a|u)p(u|s)$. To simplify the estimation of $p(u|s)$, we consider that it is uniform over the stations visited by the user in the frequency band of the activity.

We build the representation of each station for each frequency band as the sum of all the activities of all the users authentication at this station weighted by the overall usage in this frequency band. This gives us a set of three representation matrices for the station, one per frequency band, a row of which representing a particular station. As we extract 100 atoms per frequency band and there are around 300 stations in our problem, these matrices have small dimensions.

We use a multi-scale clustering algorithm, similar to what [25], to cluster together stations into coherent groups of behaviors. Since the matrices are small (300×100), the clustering is fast (a couple of seconds) on a typical computer. It is stable with respect to the initialization. For simplicity, we present the result using 5 clusters to capture macroscopic groups of behaviors in the network. In order to evaluate our approach, we compare NMF-clustering results with a basic k-means learn over raw station logs (without taking into account user profiles).

Figure 7 illustrates the result of the k-means procedure. Associated prototypes are shown on Fig. 6. Results are very hard to interpret: classically, a

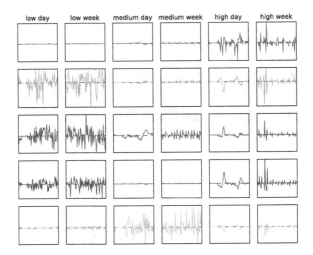

Fig. 6. Prototypes associated to each cluster from the k-means procedure. As all prototypes were very close, we choose to plot the differences between the prototype and the global mean.

main cluster corresponds to standard behavior while smaller classes model what appear as random variations around the first prototype.

Figure 8 represents the NMF clustered map of the subway stations in Paris where each station is colored (and shaped) according to its cluster. Some geographical patterns clearly emerge. There are two inner clusters in the center, one belt-like cluster around this center and the separation of the western and eastern sub-urban regions around Paris. It is interesting to compare this clustering to the sociological geography of the city. First the touristic center of Paris with *Champs Élysées* and *Concorde*, the Louvre Museum, the Garnier Opera, *Notre Dame* and the *Sacré Cœur* are in the same cluster. Second, the belt-shaped cluster, here in yellow squares, corresponds to the limits between Paris and the surrounding cities. The city limits are marked by *Porte* (gates in French) as a reminder of the gates piercing the old fortified walls of the medieval Paris. And last but not least, the clustering opposes the posh western sub-urban regions of Paris to the relatively poorer eastern sub-urban regions: the distinction, based on temporal patterns, is interesting as the users might have the same patterns since they are at the same distance from the city center. So the distinction goes beyond the simple geographic explanation and touches a sociological repartition of work hours.

The NMF has provided centroids composed of one pattern per frequency band. For any given cluster, the high frequency band comes from the usage pattern of people that frequently check-in one of the stations meaning they either live or work there. The low frequency band corresponds to people that hardly use the station (less than once every ten days) and is basically composed of evening events, meals or sleepovers. The medium band corresponds to everything in between.

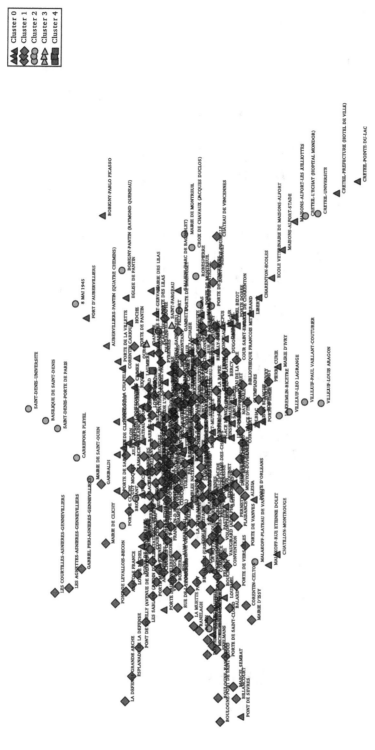

Fig. 7. Map of Paris subway stations with colors coming from the basic k-means clustering, without taking into account user profiles.

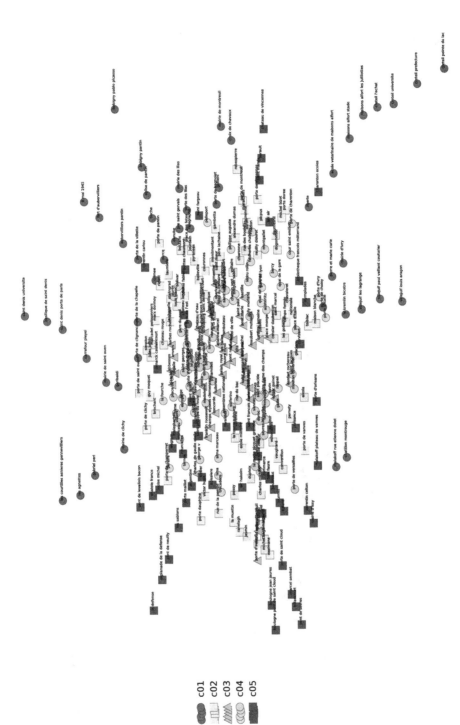

Fig. 8. Map of Paris subway stations with colors coming from our NMF clustering.

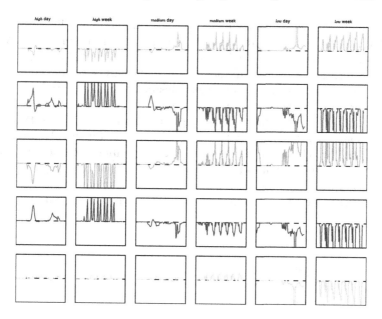

Fig. 9. Difference to the average behavior per multi-instance centroids. Juxtaposition of the day and week scales for the high, medium and low frequency bands respectively.

The difference between the patterns of the centroids and the average network load are represented in Fig. 9, with each centroid being colored as the corresponding cluster in Fig. 8. A positive spike means more users authenticating themselves at any station of the cluster than in average over the network. A negative spike is the opposite: a deficit of users compared to average load.

As only the check-in authentications are available to us, the high frequency band corresponds to the habit of the people living or working near the stations of the cluster while the medium and low frequency bands correspond to people check-in to travel elsewhere, meaning that they came to the stations of the cluster for some periodical activity like pubs, restaurants, theaters and so on and are now leaving.

The last line is the belt cluster that corresponds to the average behavior of standard Paris dwellers which is coherent with the fact that stations composing this cluster are at the limit of the city. The first and third lines are close one to another. The former corresponds to the touristic center of Paris which is sensible as it is characterized by a lack of check-ins in the morning for no working class is living there. Once again the temporal pattern is linked to a sociological repartition. The latter contains the big clusters corresponding to train stations. People working in sub-urban regions where the subway network is not present typically take the train to one hub and then finish their journey with the subway. This explains the main difference in the high frequency band between the two clusters: the peak around nine in the morning. It is also noticeable that these two clusters are the only ones having people departing after infrequent activities, as can be seen on the low frequency band. The second and fourth lines correspond respectively to the western and eastern sub-urban regions around Paris. As said earlier they are mainly residential areas with the western part being wealthier than the eastern one. This is confirmed by the night part of the medium and low frequency bands: in contrast to the center clusters, few people are leaving these clusters after some episodic activities. The phenomenon of a peak in the medium band correspond to the activity of leaving for work from a station that is not the main home of a user and is typical of sleepovers. Finally the commuting patterns of the two clusters is different. The model is able to extract fine grained representations and distinguish the clusters as they do not commute at the same time. Stations in poorer regions see a peak early in the morning followed by a deficit of users at the same time as the affluence peak in stations of posh neighborhoods.

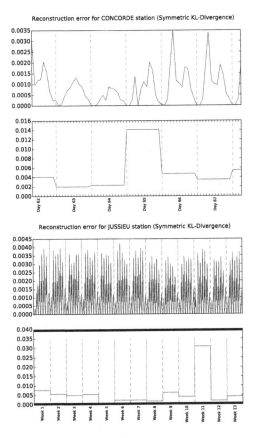

Fig. 10. Ground truth profiles and symmetrised KL divergences for 2 stations (Concorde and Jussieu) respectively aggregated by day and week.

7   Abnormality Detection

The NMF decomposition allows to reconstruct the temporal log flow of a station. We exploit this ability to build the standard weekly profile associated to every stations and we measure the distance between reconstructed data and ground truth using a symmetrised Kullback-Leibler divergence (KL). In order to improve robustness, we aggregate the results respectively over days and weeks. Figure 10 illustrates some promising results: whereas raw log flows seem roughly standard, KL measure enables us to point out clearly local abnormality at different scales.

8   Conclusion

In this work, we have proposed a new approach to urban mobility analysis introducing a machine learned based modeling that fully exploits the data available since the introduction of RFID cards in public transportation. We gave proposed a multi-scale modeling of event logs of users in order to retrieve latent activities of users. A nonnegative matrix factorization algorithm with sparsity, mono-modality and normalization constraints is used to build the set of dictionary atoms representing these activities. We have analyzed and exploited the extracted representations of the users to build stations profiles and to cluster them. We showed the interest of our formulation with respect to simpler a k-means that does not take into account user modeling. Finally we demonstrate our ability to reconstruct a temporal log distribution at the station scale as well as to detect abnormalities in the log flow.

The study of these clusters has revealed that temporal patterns are able to capture fine grained representations of behaviors from roughly aggregated noisy ticketing logs. We used here the extracted latent activities in a qualitative study. From a machine learning point of view, the extracted dictionary atoms contains meaningful high-level information that can be exploited further to jointly characterize users and locations.

One main perspective of this work concerns the design of an embedded model able to model all frequency scale at the same time. Such a model will not be easy to develop; indeed, preliminary works show that cascade approches mainly focus on frequent events without describing rare phenomena.

References

1. Berry, M.W., Browne, M., Langville, A.N., Pauca, V.P., Plemmons, R.J.: Algorithms and applications for approximate nonnegative matrix factorization. In: Computational Statistics and Data Analysis (2007)
2. Black, J.A., Paez, A., Suthanaya, P.A.: Sustainable urban transportation: performance indicators and some analytical approaches. J. Urban Plann. Dev. **128**(4), 184–209 (2002)
3. Brockmann, D., Hufnagel, L., Geisel, T.: The scaling laws of human travel. Nature **439**, 462–465 (2006)

4. Ceapa, I., Smith, C., Capra, L.: Avoiding the crowds: understanding tube station congestion patterns from trip data. In: ACM SIGKDD (2012)

5. Foell, S., Kortuem, G., Rawassizadeh, R., Phithakkitnukoon, S., Veloso, M., Bento, C.: Mining temporal patterns of transport behaviour for predicting future transport usage. In: Proceedings of the 2013 ACM Conference on Pervasive and Ubiquitous Computing Adjunct Publication (2013)

6. Golias, J.C.: Analysis of traffic corridor impacts from the introduction of the new athens metro system. J. Transp. Geogr. **10**(2), 91–97 (2002)

7. Golub, G.H., Van Loan, C.F.: Matrix Computations, 3rd edn. Johns Hopkins University Press, Baltimore (1996)

8. Gonzalez, M.C., Hidalgo, C.A., Barabasi, A.L.: Understanding individual human mobility patterns. Nature **453**, 779–782 (2008)

9. Hegde, C., Baraniuk, R.G.: Sampling and recovery of pulse streams. IEEE Trans. Signal Process. **59**(4), 1505–1517 (2011)

10. Herring, R.J.: Real-time traffic modeling and estimation with streaming probe data using machine learning. Ph.D. thesis, University of California, Berkeley (2010)

11. Hoyer, P.O.: Non-negative matrix factorization with sparseness constraints. JMLR **5**, 1457–1469 (2004)

12. Lee, D.D., Seung, H.S.: Algorithms for non-negative matrix factorization. In: NIPS (2000)

13. Liebig, T., Xu, Z., May, M., Wrobel, S.: Pedestrian quantity estimation with trajectory patterns. In: Flach, P.A., De Bie, T., Cristianini, N. (eds.) ECML PKDD 2012, Part II. LNCS, vol. 7524, pp. 629–643. Springer, Heidelberg (2012)

14. Liu, W., Zheng, Y., Chawla, S., Yuan, J., Xing, X.: Discovering spatio-temporal causal interactions in traffic data streams. In: ACM SIGKDD (2011)

15. Louail, T., Lenormand, M., Cantú, O.G., Picornell, M., Herranz, R., Frias-Martinez, E., Ramasco, J.J., Barthelemy, M.: From mobile phone data to the spatial structure of cities (2014). arXiv preprint arXiv:1401.4540

16. Peng, C., Jin, X., Wong, K.C., Shi, M., Liò, P.: Collective human mobility pattern from taxi trips in urban area. PLoS ONE **7**(4), e34487 (2012)

17. Randriamanamihaga, A., Côme, E., Oukhellou, L., Govaert, G.: Clustering the vélib origin-destinations flows by means of poisson mixture models. In: ESANN 2013 (2013)

18. Shahnaz, F., Berry, M.W., Pauca, V., Plemmons, R.J.: Document clustering using nonnegative matrix factorization. Inf. Process. Manag. **42**(2), 373–386 (2006)

19. Song, C., Qu, Z., Blumm, N., Barabási, A.L.: Limits of predictability in human mobility. Science **327**, 1018–1021 (2010)

20. Vincent, E., Bertin, N., Badeau, R.: Harmonic and inharmonic nonnegative matrix factorization for polyphonic pitch transcription. In: IEEE ICASSP (2008)

21. Wang, B., Plumbley, M.D.: Musical audio stream separation by non-negative matrix factorization. In: Proceedings of the DMRN Summer Conference (2005)

22. Wang, D., Pedreschi, D., Song, C., Giannotti, F., Barabási, A.L.: Human mobility, social ties, and link prediction. In: ACM SIGKDD (2011)

23. Wang, F., Lee, N., Hu, J., Sun, J., Ebadollahi, S.: Towards heterogeneous temporal clinical event pattern discovery: a convolutional approach. In: ACM SIGKDD (2012)

24. Zafeiriou, S., Tefas, A., Buciu, I., Pitas, I.: Exploiting discriminant information in nonnegative matrix factorization with application to frontal face verification. IEEE TNN **17**(3), 683–695 (2006)

25. Zhang, M.L., Zhou, Z.H.: Multi-instance clustering with applications to multi-instance prediction. Appl. Intell. **31**(1), 47–68 (2009)

Context-Aware Location Prediction

Roni Bar-David and Mark Last[(✉)]

Department of Information Systems Engineering,
Ben-Gurion University of the Negev, 84105 Beersheba, Israel
ronil2@gmail.com, mlast@bgu.ac.il

Abstract. Predicting the future location of mobile objects has become an important and challenging problem. With the widespread use of mobile devices, applications of location prediction include location-based services, resource allocation, handoff management in cellular networks, animal migration research, and weather forecasting. Most current techniques try to predict the next location of moving objects such as vehicles, people or animals, based on their movement history alone. However, ignoring the dynamic nature of mobile behavior may yield inaccurate predictions, at least part of the time. Analyzing movement in its context and choosing the best movement pattern by the current situation, can reduce some of the errors and improve prediction accuracy. In this chapter, we present a context-aware location prediction algorithm that utilizes various types of context information to predict future location of vehicles. We use five contextual features related to either the object environment or its current movement data: current location; object velocity; day of the week; weather conditions; and traffic congestion in the area. Our algorithm incorporates these context features into its trajectory-clustering phase as well as in its location prediction phase. We evaluate the proposed algorithm using two real-world GPS trajectory datasets. The experimental results demonstrate that the context-aware approach can significantly improve the accuracy of location predictions.

1 Introduction

In recent years, ever more data is available about the location of moving objects and their usage context. Playing no small part in this growth is the increasing popularity of mobile devices equipped with GPS receivers that enable users to track their current locations. These devices have now emerged as computing platforms enabling users to consume location-aware services from Internet websites dedicated to life logging, sports activities, travel experiences, geo-tagged photos, etc. In addition, many users have started sharing their outdoor movements using social networking sites such as Facebook and Twitter, which have now added location as an integral feature. The interaction of users with such services inevitably leaves some digital traces about their movements in the device logs and on various web servers, which provide a rich source of information about people's preferences and activities with regard to their location. This provides an opportunity to collect both spatio-temporal and contextual data, and in turn, to develop innovative methods for analyzing the movement of mobile objects.

We assume that the behavior of moving objects can be learned from their historical data. In any particular environment, people do not move randomly but rather tend to

M. Atzmueller et al. (Eds.): MSM, MUSE, SenseML 2014, LNAI 9546, pp. 165–185, 2016.
DOI: 10.1007/978-3-319-29009-6_9

follow common paths, specific repetitive trajectories that correspond to their intentions. Thus, predicting their future location is possible by analyzing their movement history. Location prediction models can enhance many different applications, such as location-aware services [1], mobile recommendation systems [2], animal migration research, weather forecasting [3], handoff management in mobile networks [4], mobile user roaming [5], etc. While a substantial amount of research has already been performed in the area of location prediction, most existing approaches build upon the most common movement patterns without taking into account any additional contextual information. This severely limits their applicability since in practice the movement is usually free and uncertain. In many situations, it may not be sufficient to consider only the movement history, since the probability of a future location does not only depend on the number of times it was visited before but also on the object's current context. For example, in different situations, a driver may choose different routes leading to the same destination.

The importance of contextual information has been recognized by researchers and practitioners in many disciplines. Context-aware applications are able to identify and react to real-world situations better than applications that do not consider this sort of information. Context information can help systems to ease the interaction with the user and sometimes even allow a device to be used in a way that was not intended by design. In this chapter, we show that relevant contextual information is important for providing accurate location predictions.

In this work, we represent movement patterns of vehicles as sequences of spatio-temporal intervals, where each interval is a minimal bounding box (MBB). We enhance the standard MBB-based representation of movement patterns described in [6] with a context profile. The extended MBB represents a geographic area traversed by a vehicle at a certain time, which also has a semantic meaning. The context profile turns an MBB into a segment that represents a set of data points not only close in time and space, but also having the same or similar values of contextual features. We extract several types of contextual features for each MBB from its summarized data points. These contextual features include the vehicle's average velocity, day of the week, weather conditions for the area where the vehicle is moving, and traffic conditions in the movement region. We use a context-aware similarity measure to cluster similar trajectories with different types of contextual data. We present a new algorithm for location prediction that utilizes the contextual features and overcomes the limitation of context-free algorithms by choosing the relevant movement pattern according to the current context. Finally, we evaluate the prediction performance and show that our context-aware prediction model can provide more accurate predictions than a context-free model. The main contribution of this work is the use of spatio-temporal data mining techniques in combination with context information in order to enhance the accuracy of location prediction methods.

2 Related Work

2.1 Location Prediction Methods

Predicting the future location of a given mobile object is an interesting and challenging problem in data mining. Monreale et al. [7] propose the WhereNext system that is based on association rules, and define a trajectory as an ordered sequence of locations. Given a trajectory of a moving object, it selects the best matching association rule and uses it for prediction. Trajectory patterns are sequences of spatial regions that emerge as frequently visited in a given order. Motion functions represent the trajectory of an object based on its recent movements, and indicate location changes by updating the function parameters, instead of recording object locations at individual timestamps. The most common functions are linear models that assume an object follows linear movements [8]. Given an object's current location, time and velocity the object's location at a future time is estimated. Unfortunately, the simple linear model has a limited applicability due to the enormous diversity of motion types. There are non-linear models that capture the object's movements by sophisticated mathematical formulas. Consequently, they have a higher prediction accuracy than the linear models [9]. The recursive motion function is the most accurate prediction method among motion functions. It predicts an object's location at a specific time in relation to those of the recent past. It can express unknown complex movements that cannot be represented in an obvious manner. Jeung et al. [10] propose a hybrid prediction algorithm that takes advantage of an object's pattern information, as well as its motion function. This algorithm mines an object's trajectory patterns that follow the same routes over regular time intervals. When there is no pattern candidate in the prediction process, it calls a recursive motion function to answer the query.

Nizetic et al. [11] use Markov Chain models to address the problem of short-term location prediction. This paper describes a Hidden Markov Model for routing management in mobile networks. A Hidden Markov Model assumes that the current object state is conditionally dependent on its previous k states. Each movement between locations corresponds to a transition between states and it is assigned a certain probability. In a Hidden Markov Model, the underlying states are not directly observable, and there is only access to an observable set of symbols that are probabilistically related to the hidden states.

Another use of Markov models for location prediction is presented by Ashbrook and Starner [12]. The proposed system converts GPS data into sequences of meaningful *locations*, which are defined as geo-spatial clusters of places where the users stopped for at least 10 min. Given the current location context, the Markov model provides the probability of transition to every other location. Extending the proposed methodology to other sources of context is mentioned as part of future research.

Liao et al. [13] apply a hierarchical Markov model for learning and inferring a user's daily movements from GPS data. All locations are constrained to be located on a street map. The location of the person at time k is assumed to depend on his previous location, l_{k-1}, the velocity, v_k, and the vertex transition, τ_k. The hierarchical prediction model takes into account a person's goals (trip destinations), trip segments (transition probabilities on the intersections graph), and novelty of his movement behavior.

Clustering is an unsupervised method of learning, which reveals the overall distribution pattern and interesting correlations in datasets. A cluster is a set of data objects with similar characteristics that can be collectively treated as one group. Han and Yang [12] introduce the concept of moving micro-clusters. A micro-cluster indicates a group of objects that are not only close to each other in a specific time, but also likely to move together for a while. In order to keep the high quality of clustering, these micro-clusters may split and then reorganize. A moving micro-cluster can be viewed as one moving object, and its center can be treated as its location. Elnekave et al. [13] propose a segmentation algorithm for representing a periodic spatio-temporal trajectory as a compact set of MBBs. An MBB represents a spatio-temporal interval bounded by limits of time and location. This structure allows summarizing close data-points, such that only the minimum and maximum values of the spatial and time coordinates are recorded. The authors define data-amount-based similarity between trajectories according to proximity of trajectories in time and space.

2.2   Spatio-Temporal Context Models

Context is a general concept describing the conditions where a process occurs. The term has been used in many ways across different research disciplines. Several definitions of the term "context" have evolved over years. The earliest definition of context-aware computing was suggested by [14] who regard the term as referring to the user's location, the identity of people near the user, nearby objects, and the changes in these elements. Schmidt et al. [15] enriched context definition to include additional user information, such as habits, emotional state, position shared with others, and social interactions. Chen and Kotz [16] further extended context to involve more environmental information, and distinguished between active context that influences the behavior of an application and passive context that can be relevant to the application but not critical for it. Sanchez et al. [17] explained the distinction between raw sensor data and context information. Raw data is unprocessed and retrieved directly from the data source, whereas context information is generated by processing, validating and adding metadata to the raw data. For example, GPS coordinates are raw data that can represent a geographical location as context information.

Context awareness is defined as the ability to identify and react to situations in the real world by taking advantage of contextual information [16]. Making use of relevant context information from a variety of sources to provide suitable services or task-dependent information is one of the main challenges in ubiquitous computing. Zhang et al. [18] define ubiquitous computing as a paradigm shift where technologies become virtually invisible or transparent and yet pervasively affect every aspect of our lives. The authors proposed a reference framework to identify key functionalities of context-awareness, and assisting ubiquitous media applications in discovering contextual information and adapting their behavior accordingly. Another context-aware model that supports managing of ubiquitous applications is described by Lopes et al. [19]. It includes modules for context acquisition from distributed sources, context storage and actuation in the environment, and context processing for notifying the current state.

Context information is used in the location prediction literature mostly with respect to identity, location and time. Zheng et al. [20] demonstrate the use of user location as context in their supervised learning approach for detecting spatial trajectory patterns. They mined travel sequences of multiple users from GPS trajectories, and found interesting places in a certain region from the location history. An interesting location is defined using a stay point, a geographic region where the user stayed over a certain time interval. This model uses other people's travel experiences in order to improve travel recommendations. Gao et al. [21] propose to capture the spatio-temporal context of user check-in behavior in order to improve a location prediction model. They consider temporal periodic patterns in addition to spatial historical trajectories for computing the probability of the next visit in a location during a certain time interval. The location and time probabilities restrain and complement each other in the form of spatio-temporal context. Appling smoothing techniques they found that a user mostly checks-in at a location during a specific time interval and rarely visits during other time intervals. Mazhelis et al. [22] propose incorporating temporal context dimensions such as time of the day in a route recognition system. They designed a real-time personal route recognition system that uses context information to predict the route destination. Instead of comparing the current driving trajectory against the trajectories observed in the past and select the most similar route, they incorporated the time of day as context, and adjusted the route recognition output accordingly. They found that the trajectories depend on the time of day feature, even at the beginning of a trip, that typically starts at the same departure point (e.g., home) and creates overlapping trajectories. Assam and Seidl [25] have recently presented a location clustering and prediction technique based on a novel context-based similarity measure for locations. This new measure uses a combination of two similarity measures, namely *user similarity* and *time similarity*. The underlying assumption is that the same group of people would meet together at the same time in the same *location context* (activities such as having lunch or attending a daily business meeting).

In our literature survey, we could not find any location prediction solutions that utilize context information from multiple data sources. Current techniques often focus on only one type of spatial or temporal context, and do not consider sudden changes in the movement and the environment. The attempts to integrate both spatial and temporal context information suffer from the overfitting problem due to the large number of spatio-temporal trajectory patterns [21]. In order to cope with the dynamic nature of movement, which may cause a user's behavior to be inconsistent with his historical movement patterns, we are interested in context features that affect the choice of a movement trajectory. Rather than trying to incorporate the context information in the task model, the existing context-aware models treat the context model and the task separately. In this chapter, we propose a convenient representation of the context information, which can be incorporated in the movement pattern representation. Our goal is to integrate into one prediction model both the context features extracted from the historical movement data and the context features available from external sources, like web services. The following contextual features are used here for the first time: day of the week, weather conditions, and traffic congestion.

3 Proposed Methodology

Our approach consists of three main steps: (1) extracting periodic trajectories together with the context information of the mobile object; (2) inducing context-aware models that describe the object's movement patterns; and (3) predicting the future location of an object based on the two previous steps and the current context. The initial operations for extracting context features and mining the movement patterns of a moving object (Steps 1 and 2) can be performed off-line. The last step (Step 3) involves a context-aware location prediction algorithm that should be implemented on-line. In the prediction phase, we incorporate context information by selecting only movement patterns that satisfy the context information criteria for a given time. This approach is based on the concept that each pattern has its own particular context. For example, if it is currently raining, only the movement patterns that were created under raining conditions will be left in the candidate set for prediction. By using contextual information in real time, we are able to choose the most likely movement pattern rather than relying on the most common one. Figure 1 presents a block diagram describing the proposed methodology for context-aware location prediction.

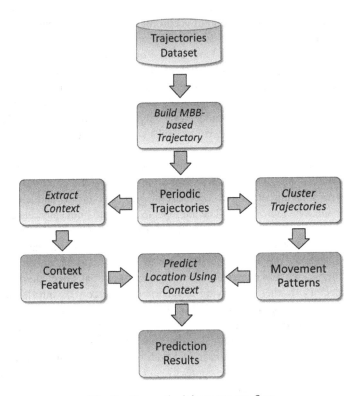

Fig. 1. The methodology process flow

3.1   Extracting Contextual Information

We introduce a unique representation of an area and its contextual features, as a combination of a minimal bounding box (MBB) and a context profile. The extended MBB represents a spatio-temporal interval as well as its related context features. It can be regarded as a geographic area traversed by an object at a certain time, which also has a semantic meaning. Formally, MBB has the following properties:

$$i.t_{min} = \min(\forall p \in i, p.t_{min}), \; i.t_{max} = \max(\forall p \in i, p.t_{max}) \tag{1}$$

$$i.x_{min} = \min(\forall p \in i, p.x_{min}), \; i.x_{max} = \max(\forall p \in i, p.x_{max}) \tag{2}$$

$$i.y_{min} = \min(\forall p \in i, p.y_{min}), \; i.y_{max} = \max(\forall p \in i, p.y_{max}) \tag{3}$$

where i represents an MBB; p represents aata point in a box; x and y are spatial coordinates; and t_{min} and t_{max} are the object's minimum and maximum times in the area. This allows us to summarize close data points into one MBB, and represent it by these six elements rather than by the multiple original data points. We consider context to be represented by a context profile related to both an area and a mobile object's current state. Context features may be related to the environment surrounding a mobile object or to the mobile object movement itself. A context profile turns an MBB into a segment that represents a set of data points not only close in time and space, but also having the same or similar values of contextual features. A context feature has an identifier, a type and a value. Formally:

Context Profile = (i, amount, location, velocity, day of the week, weather, traffic congestion)
$$\tag{4}$$

where i identifies the MBB; and the other properties represent the aggregated feature values. We enhance the MBB representation to support additional context features with the following method:

$$MBB.cf = \text{aggregation}(\forall p \in MBB, p.cf) \tag{5}$$

where *cf* stands for a context feature that is being aggregated; and p represents a point member in an MBB. In case of day of the week and traffic congestion, the aggregation is straightforward as data points with different labels are assigned to different MBBs automatically and thus, all data points in a given MBB will have the same contextual value. The weather feature is aggregated to the most frequent contextual value in a given MBB. Numeric features (location and velocity) are aggregated to the mean value of all data points in a given MBB.

This compact and unique representation of an MBB allows us to obtain accurate and timely results in subsequent mining stages, and to incorporate the context information in the trajectory-clustering phase. Several context features may influence the mobile object movement. We are looking for occurrences of certain context characteristics that change the movement habits. Awareness of such events can help us choose the right movement pattern, and help to improve future location predictions. A diagram describing the different types of context is shown in Fig. 2. A detailed description of each context feature is provided below.

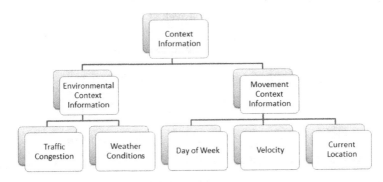

Fig. 2. Different context types

Amount. This is the baseline feature that represents the number of data points that are summarized by a given MBB. It is calculated as the number of times the area has been accessed. The more times an object visits a particular MBB, the more important it becomes. The prediction algorithm chooses the most common MBB within the specified time bounds based on the maximal data amount property. Formally:

$$MBB.amount = \text{count}(\forall p \in MBB, 1). \qquad (6)$$

Current Location. The current location of every point is recorded in the dataset using Cartesian coordinates. Though it is the simplest feature to extract, it is very powerful because it allows us to discard common, yet infeasible, patterns. We can correct unreasonable predictions by choosing other movement patterns. For instance, if a mobile object usually travels from home to work every morning, and we find it near the other side of town, we should change our prediction. We calculate the center point of each candidate MBB, and choose the closest one to the current object position. We use the Haversine distance formula [23] to calculate the distance between two points. At small scales, this formula acts exactly like the Euclidean distance formula and can provide a good approximation of road distance. At large scales, however, distances along the surface of earth are more curved and therefore using the straight-line distance is less suitable. The Haversine formula uses the point longitude and latitude and calculates the shortest distance (in km) over the earth's surface by the following formula:

$$H.distance = \text{R} \cdot 2 \cdot \text{atan2}(\sqrt{\alpha} \cdot \sqrt{1 - \alpha}) \qquad (7)$$

$$\alpha = \sin^2(\frac{\Delta latitude}{2}) + \cos(lat1) \cdot \cos(lat2) \cdot \sin^2(\frac{\Delta longitude}{2}) \qquad (8)$$

where R is the earth's radius that has a fixed value of 6,371 km; $atan2()$ is a trigonometric function; and Δ is the subtraction difference between the points longitude and latitude values.

Velocity. Moving at a certain velocity may imply the next destination of the object. We assume that different velocities can represent different contexts, and if an object's velocity is higher than average, it is an indication that its destination is different than the usual one and vice versa. For example, driving at a high velocity near the work area may imply that the destination is not the office. We assume that each mobile object has its actual velocity recorded at every data point. If it is missing, we can compute it from the spatio-temporal trajectory by dividing the distance travelled between two consecutive points by the time between the two measurements. We estimate the average velocity at each data point as the average velocity over the last five points. The average velocity of a given MBB is calculated over its all data points. The prediction algorithm chooses only MBBs with the average velocity similar to the object current velocity.

Day of the Week. The current time is recorded as a timestamp for each entry in the dataset. Thus, we know the current weekday and try to find typical routes for the regular weekdays and the weekend. In order to model this contextual feature, the week is divided into two periods: the regular weekdays and the weekend. Each point or MBB is associated with one of these periods, according to the trajectory time. We assume that the mobile object movement depends on the day of the week context. For instance, on a regular weekday a person may drive to her/his office in the morning, whereas on the weekend he will probably choose a different, recreational destination. The prediction algorithm chooses MBBs according to the current day of the object's movement.

Weather. Weather conditions may affect the movement of every object. For example, snow can prevent drivers from accessing certain areas whereas downpours can cause people to assemble in the mall instead of a public open space. Since the movement data does not contain information about weather, we should obtain this information from external data sources. For example, we may use an online weather Web service called Weather Underground (http://www.wunderground.com/) that publishes weather data for many locations on an hourly basis and describes weather conditions with different numeric and Boolean variables. We divide the object movement area into a grid, consisting of squares with a certain parameter size. We then query the site for each new point, and save the weather information in order to reuse it as long as the location does not exceed a threshold value. We refer only to the three weather conditions that may influence the object movement, namely foggy, rain, and snow. Any other type of weather condition, such as clear, cloudy, hot, cold, etc. is ignored and labeled as the "other" category. We retrieve weather conditions in an MBB by using the mean coordinates and the median timestamp. An MBB weather is defined as the most frequent category in its area and within its time bounds.

Traffic Congestion. Traffic congestion affects every mobile object by slowing its normal speed and causing a change in its routine movement habits. Since the movement data does not contain explicit information about traffic congestion, we implicitly infer this data. Using an external Web service or a community-based application like Waze, we can locate areas that are prone to traffic congestion by searching for certain events of decreasing velocity in their vicinity. Traffic congestion is characterized by a significant difference between the current speed and the permitted speed on the road. When an object is moving much slower than the allowed speed in the area, we infer that a traffic congestion event takes place. In the case when we do not have access to the permitted velocity on the road, we look for areas in which the velocity decreases for a certain distance and certain time. We use the segmentation algorithm described in Sect. 3.2 to extract traffic congestion areas containing the mentioned events, with duration above 10 min and distance above 200 meters. We mark these MBBs with a traffic congestion label. The clustering algorithm described in Sect. 3.3 finds areas where these kinds of events occur repeatedly. Not every decreasing velocity event corresponds to traffic congestion. The decreased velocity may also occur close to a fueling or a parking area. The MBBs marked with a traffic congestion label are areas containing decreasing velocity events that occur closely, both in space and in time. Thus, occasional low velocity events that occur closely in space but at different times will be regarded as noise.

In order to verify our traffic congestion findings, we may use an external web service such as Google Places that suggests points of interest, which are close to a given location. Points of interest include culturally important places and commonly frequented public areas, like shopping malls and restaurants. Google Places supports 126 metadata types to describe points of interest. We refer only to the following three categories: institutions (including religious, health and services), entertainment (including shopping, parks and food) and transportation (including bus/train stations and intersections). We search points of interest in the one-kilometer radius of every MBB center point. If an MBB is located near highway intersections around the city, or on major streets, it is more likely that traffic congestion can be expected. In the prediction phase, when we detect a traffic slowdown event, we adjust the time bounds respectively. Thus, the prediction algorithm can choose only feasible MBBs, which are close to the current object location and are characterized by a similar average velocity.

Combined Context. This method integrates all the above context features together into one model. We calculate the context similarities for each candidate MBB within the prediction time bounds. The final similarity measurement assigns an equal weight to all five context features and thus utilizes their overall influence on the movement trajectory.

3.2   Building MBB-Based Trajectory

We present an incremental segmentation algorithm (denoted as *Algorithm I*) that summarizes the movements of each mobile object into periodic trajectories while extracting the desirable context features. We are looking for trajectories that continually reoccur according to some temporal pattern in a certain context. The algorithm inputs are a spatio-temporal dataset that contains GPS log entries, the available context feature, and a scaling parameter to determine MBB bounds. The threshold values are calculated by multiplying the data-points standard deviation in each dimension, with the scaling parameter. The larger the threshold is, the more summarized the trajectories are.

The outputs of this phase are an object's periodic trajectories represented as a list of MBBs. These segmented trajectories become the input of the next, clustering phase.

Algorithm I. Building an Object's Context-Aware Trajectory
1. *trajectory*.addNewMBB(*point*$_0$) // add the first MBB
2. *i* ← 1
3. **For Each** *point*$_i$ ∈ *dataset*
4.   If *point*$_i$. *date* = *point*$_{i-1}$. *date*
5.   If |*point*$_i$.x - *trajectory*.lastMBB.maxX| < *threshold*$_X$ **And**
6. |*point*$_i$.y - *trajectory*.lastMBB.maxY| < *threshold*$_y$ **And**
7. |*point*$_i$.time - *trajectory*.lastMBB.maxTime| < *threshold*$_{Time}$ **And**
8.   *point*$_i$.cf= *trajectory*.lastMBB.cf
9.   *trajectory*.lastMBB.addPoint(*point*$_i$)
10.   **Else**
11.   *trajectory*.addNewMBB(*point*$_i$)
12.   **Else**
13.   *result*.add(*trajectory*)
14.   *trajectory* ← null
15.   *trajectory*.addNewMBB(*point*$_i$) // start new periodic trajectory
16. **Return** *result*

In lines 1–4 of Algorithm I, we process each incoming data point. In lines 5–9, as long as the point has the same context and is within the MBB bounds, we insert it into the current MBB. In lines 10–11, we decide that the point stretches the MBB beyond the threshold and create a new MBB. The *addNewMBB* function initializes a new MBB in the trajectory, and initializes its bounds and context according to the data point. The *addPoint* function increments the MBB amount property and updates the existing MBB boundaries based on the new point coordinates.

3.3   Context-Aware Similarity Measure

We define a similarity measure that takes into account the context features in the trajectory clustering algorithm. We incorporate the context into the model by increasing the similarity for MBBs in the same context. For example, if one MBB has a raining weather condition, all MBBs with the same weather condition will get bonuses towards their similarities with the current MBB. All other MBBs will remain in the candidate set without decreasing their similarity. We define the similarity between two trajectories as the sum of similarities between their corresponding MBBs:

$$TrajectorySim(T_1, T_2) = \frac{\sum_{i=1}^{n} \sum_{j=1}^{m} MBBsim\left(T_1^i, T_2^j\right)}{n \times m} \qquad (9)$$

where T_1 is the first trajectory; T_2 is the second trajectory; T_1^i is the i MBB segment of the first trajectory; T_2^j is the j MBB segment of the second trajectory; n is the amount of MBBs in the first trajectory; and m is the amount of MBBs in the second trajectory.

Our context-based similarity is calculated as the product of three factors: (1) the overlapping time of two MBBs; (2) the similarity between the data point amounts summarized within the two MBBs; and (3) the minimal distance between the two MBB features. Formally:

$$MBBsim(MBB_1, MBB_2) = |t_m - t_n| \times \#points \times minContextDist(MBB_1, MBB_2) \quad (10)$$

where t_m is the time when the two MBBs start to overlap; and t_n is the time when its overlapping ends. The greater the overlapping time between the two MBBs, the higher the similarity between them. The *#points* parameter is the minimum amount of data points within the two compared MBBs. The more data points are included in both of the compared MBBs, the stronger support we have for their similarity. The minimal context distance between two MBBs, consists of the sum of minimal distances of every feature in the MBBs. The minimal distance between two features is defined as the distance in that dimension or as zero if the two MBBs overlap. Formally:

$$minContextDist(MBB_1, MBB_2) = \sum_{i=1}^{n} minCFDist(MBB_1.cf_i, MBB_2.cf_i) \quad (11)$$

$$minCFDist(MBB_1, MBB_2) = \max(0, \max(MBB_1.cf_{min}, MBB_2.cf_{min}) \\ - \min(MBB_1.cf_{max}, MBB_2.cf_{max})) \quad (12)$$

where n is the number of context features we are measuring; and cf is the value of the feature i. We combine two distance values: a spatial distance and a context distance using a weighted average approach. For numeric features such as velocity, the distance is defined using this formula: $|v_1 - v_2|/(v_1 + v_2)$ where v_1, v_2 are feature values. When there is no difference between the values, the distance is zero, otherwise it is the normalized distance between the values, so the distances have a comparable scale. For nominal features such as weather and day of the week, the distance is defined as zero if the context values are the same and as one, otherwise.

We enhanced the k-means incremental clustering algorithm described in [6] to use this similarity measurement in order to induce context-aware movement patterns from periodic trajectories. The algorithm gets as an input a set of periodic trajectories, a number of clusters, iteration bound, and a context feature. It initializes the centroids with the trajectories that are first to arrive. Then, each subsequent trajectory is inserted into its nearest centroid cluster. We update the centroids by adding each MBB into the trajectory that represents the cluster's centroid. We stop clustering when there is no change or when we reach the iteration bound. The algorithm output is the trajectory cluster's centroids, which represent the movement patterns.

3.4   Context-Aware Location Prediction

Algorithm II is aimed at predicting the location of a mobile object in the next time slot based on the context-aware movement patterns we mined from the trajectories history, along with the current contextual features. The prediction algorithm filters MBBs by context, leaving only those that have the same context as the currently observed instance. For example, if it is currently raining, only the MBBs that are labeled by the raining category will be left in the candidate set. The algorithm inputs are the next time slot (denoted as *time*), a set of cluster centroids that are the movement patterns, and a context profile *cf* that contains the current values of the defined context features. The algorithm returns the MBB's maximal and minimal coordinates as the prediction result.

Algorithm II. Predicting Mobile Object's Future Location
1. **For Each** $c_i \in centroids$
2. $j \leftarrow 0$
3. $MBB \leftarrow c_i.getMBB(j)$
4.   **While** $(MBB_{maxTime} < time)$ // proceed to relevant MBBs
5. $MBB \leftarrow c_i.getMBB(j)$
6. $j{+}{+}$
7.   **While** $(MBB_{minTime} \leq time \leq MBB_{maxTime})$ // add to candidate set
8. $MBB \leftarrow c_i.getMBB(j)$
9. $candidates.add(MBB)$
10. $j{+}{+}$
11.   **If** $MBB_{minTime} > time$ **And** $candidates.size = 0$
12.   **If** $j > 0$ // if empty set, choose the closest MBB
13. $prev \leftarrow c_i.getMBB(j-1)$
14.   **Else**
15. $prev \leftarrow c_i.lastMBB$
16.   **If** $(MBB_{MinTime} - time > time - prev_{maxTime})$
17. $candidates.add(MBB)$
18.   **Else**
19. $candidates.add(prev)$
20. $candidates.sortByAmount()$
21. **For Each** $cf_i \in cf$ // iterate context features
22.   **For Each** $MBB \in candidates$
23.   **If** $MBB_{cf} \neq cf_i$
24. $candidates.remove(MBB)$
25. $removed.add(MBB)$
26.   **If** $candidates.size = 0$
27. $candidates \leftarrow removed$ // ignore feature
28. **Return** $candidates.getMaxAmount()$

In lines 1–3 of Algorithm II, we search for MBB centroids that are within the time bounds. In lines 4–10, we add relevant MBBs to the candidate set. In lines 12–15, if no MBB matches the input time, we look for the closest MBB. The previous MBB can refer to the preceding period in the centroid, as before the first item at time 0:00 comes the last item at time 23:59. Finally, in lines 21–26, we iterate through the context profile and filter the candidates by the required features. The feature is ignored if none of the candidates is left in the candidate set. In lines 28, we choose the MBB with the highest amount of data points and return it as the prediction result.

4 Empirical Evaluation

4.1 Performance Measures

We evaluated our context-aware prediction algorithm on real-world datasets comprised of spatio-temporal trajectories of mobile objects and data from external web services. Our research hypothesis is that the context-aware algorithm should be able to predict future locations in a more accurate manner than a context-free method. Our algorithm predicts future locations in the form of MBBs rather than raw data points. In order to assess the prediction capability, we used two performance measures. The first measure is the prediction accuracy, which is the probability that the actual future location is found within the predicted MBB. Accuracy is defined as:

$$Accuracy = \frac{hits}{hits + misses}. \tag{13}$$

According to this measure, larger MBBs should have better prediction accuracy than smaller MBBs. For example, if the clustering phase produces only one large MBB, which contains all predicted locations, we will have 100 % accuracy but the prediction will not be useful for any practical purposes. On the other hand, it will be hard to build an accurate prediction model using very small MBBs. The second measure is the Mean Average Error (MAE), or the spatial distance between the real location of the object at the predicted time and the borders of the predicted area. We calculate minimal, average and maximal Euclidean distance between the actual object location and the predicted MBB's boundaries. The maximal distance is an upper boundary to the prediction error. MAE is defined as:

$$MAE = \frac{\sum maxDistance(object_{X,Y} - MBB_{X,Y})}{N} \tag{14}$$

where $object_{X,Y}$ represents the real position of the object, MBB is the prediction result, $maxDistance$ is the maximal Euclidean distance function, and N stands for the total number of predictions. Out of two predicted MBBs that both contain the real location of an object, the MAE measure would give a preference to the smaller MBB.

4.2 Experimental Datasets

We performed experiments on two real-world spatio-temporal datasets: INFATI and GEOLIFE. The INFATI dataset [24] consists of GPS log-data from 20 cars driving by family members in Aalborg, Denmark. The movement of each car was recorded for periods of between 3 weeks to 2 months during December 2000 and January 2001. When a car was moving, its GPS position was sampled every second. The GPS positions were stored in the Universal Transverse Mercator (UTM 32) format. No sampling was performed when a car was parked. The data was modified in order to allow the drivers some degree of anonymity. We referred to the cars as one group containing 20 mobile objects. The dataset contains 1.9 million records. Each record comprises GPS coordinates, time and date, velocity, and the maximum velocity allowed on the particular road where the driver was driving.

The GEOLIFE dataset [25] is a trajectory dataset that was collected mostly in Beijing, China, from 182 users, during a five-year period (2007–2012) as part of Microsoft's GEOLIFE Project. Some users carried the GPS logger for the whole period while others only carried the logger for a few weeks. The dataset as a whole recorded a broad range of users' outdoor movements, including routine matters such as going home and to work but also entertainment and sport activities, such as shopping, sightseeing, dining, hiking, and cycling. The users indicated the mode of transportation of their trajectories, such as driving, taking a bus, riding a bike and walking. We referred only to object trajectories traversed by vehicles, which were based on data collected for more than a one-month period. The dataset contains 17,621 trajectories based on a total traveling distance of about 1.2 million kilometers in the course of 48,000+ hours. These trajectories were recorded by various GPS loggers and GPS-phones at a variety of sampling rates.

The statistics of both datasets are presented in Table 1. Following the approach presented in Subsect. 3.1 above, the context information on weather conditions and places of possible traffic congestion was obtained from the Weather Underground and the Google Places web services, respectively.

Table 1. Datasets statistics

Parameter	INFATI	GEOLIFE
Mobile objects	20	62
Average dates per object	34	43
Average records per object	94,754	91,551
Total dates	679	3,293
Total records	1,895,085	6,274,996
Sampling rate	1 s	$\sim$1–5 s or $\sim$ 5–10 m

4.3   Compared Methods

In the algorithm empirical evaluation, we tried to predict the future location of mobile objects with and without each context feature. The evaluation of each context method was conducted as follows. We used all the data from every object as an independent dataset. We divided the dataset into training and testing sets, using the leave-one-out cross-validation method. In every iteration, we selected two dates as the testing set while keeping the rest of the dates as a training set. The clustering algorithm induced context-aware movement patterns from the spatio-temporal trajectories in the training set.

Afterwards, we iterated the data points from the two-day testing set, updated the context profile and predicted their future location. We used the one-tailed t-test, to test for significant difference between the following methods:

- *Baseline*. This approach, adopted from [6], is based on the frequency of visits to each area regardless of the object's specific situation. The predicted MBB is the most common MBB within the time bounds, i.e. the MBB having the maximum amount of data points.
- *Location*. This prediction method chooses the closest MBB to the current object's location out of a set of candidate MBBs. The spatial distance is calculated as the Haversine distance between the current location and the MBB centroid.
- *Velocity*. The prediction algorithm chooses only MBBs with average velocity similar to the object's current velocity. With this method, if the object is currently moving at high speed in a certain direction, the method will choose MBBs with high average speed and the same direction. If more than one candidate MBB exists, the Baseline criterion is applied.
- *Day of the Week*. The week is divided into two periods, regular weekdays and weekends, and the prediction algorithm chooses an MBB from the current day category. In case, there is more than one candidate MBB, the most common MBB is selected according to the Baseline approach.
- *Weather*. The prediction algorithm considers only MBBs with the same weather conditions as in the object area. In the clustering phase, we find the most common weather category for each MBB. In the prediction phase, we check the current weather and filter the candidate MBBs by their weather category, selecting the most common MBB as the prediction result.
- *Traffic Congestion*. We identify traffic congestion on the road as described in Sect. 3.1 and save this indication in the MBB profile. If in the prediction phase we detect a traffic slowdown in the mobile object area, we adjust the prediction time bounds respectively so that the algorithm can choose an MBB that is closer to its current position.
- *Combined Context*. The combined context method integrates all of the above context features into one prediction model. We calculate the context similarities for each possible MBB candidate within the time bounds while assigning equal weights to all five context features.

4.4 Evaluation Results

We evaluated our algorithm with different sizes of the training and testing sets. The last one-third of the dates in the dataset was selected to be the testing set. Then we let the algorithm learn the movement of a mobile object as we gradually increased the training set size from zero to two-thirds of the dates in the dataset. We compared the baseline method with the combined context method.

As can be seen in Figs. 3 and 4, location prediction accuracy is affected by the training set size with and without context information. In the INFATI dataset, the accuracy with the maximum training set size, without context information, is 0.68, deteriorating to 0.32 with only 10 % of the dataset. In the GEOLIFE dataset, when the training set consists of 10 % of the dataset, the algorithm accuracy is 0.41. As we increase the size of the training set, the accuracy reaches 0.84. As the training set size grows, the training examples tend to be more representative of the movements in a given area. Therefore, the prediction accuracy tends to be higher.

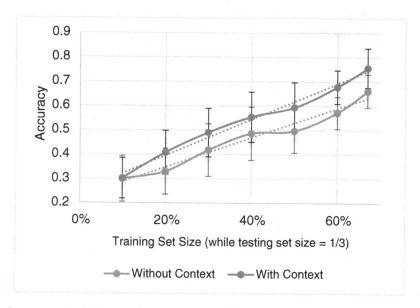

Fig. 3. Accuracy by different training set sizes (INFATI dataset). Context = Combined context.

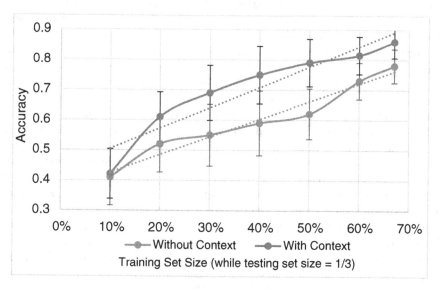

Fig. 4. Accuracy by different training set sizes (GEOLIFE dataset). Context = Combined context.

Using context information can help reduce the minimal period of movement data collection prior to performing effective location predictions. In the INFATI dataset, the accuracy becomes reasonable (above 0.6) with a training set size of 66 % of the dataset. The same level of accuracy is achieved by the context-aware method with only 50 % of the dataset. In the GEOLIFE dataset, the best accuracy result, without context information, is 0.78. The same result is achieved by the context-aware algorithm with only 50 % of the training set size.

The results of each method are summarized in Table 2.

Table 2. Accuracy and MAE prediction results

	INFATI		GEOLIFE	
	Accuracy	MAE	Accuracy	MAE
Baseline	0.738	1,067	0.814	683
Location	0.769*	943	0.830*	584
Velocity	0.799*	921	0.835*	622
Day of the week	0.731	1,239	0.801	751
Weather	0.738	1,096	0.820	641
Traffic congestion	0.812*	898	0.837*	523
Combined context	0.823*	850	0.883*	474

In both datasets, most of the context information methods outperform the baseline method. In the INFATI dataset, when using the location context method which ignores distant predictions, the accuracy increased to 0.769 (p-value < 0.001). When using the velocity context method, the accuracy was even better 0.799 (p-value < 0.0001). In the GEOLIFE dataset, the location context method increased the accuracy to 0.830 (p-value < 0.001). When using the velocity context method, the accuracy was somewhat better 0.835 (p-value < 0.001). The day of the week method did not perform better than the baseline method. The weather method has a slightly better, yet not statistically significant, prediction accuracy of 0.820 (p-value = 0.383). In the INFATI dataset, the weather method has the same prediction accuracy of 0.738 as the baseline method, with no significant improvement. This is probably due to the short period of the data collection. The best single context feature in both datasets is the traffic congestion method that combined location and velocity features and improved the prediction accuracy to 0.812 (p-value < 0.001) in the INFATI dataset, and to 0.837 (p-value < 0.001) in the GEOLIFE dataset. Finally, the combined context method which utilizes all the context features significantly improved prediction accuracy to 0.823 (p-value < 0.001) in the INFATI dataset, and to 0.883 (p-value < 0.001) in the GEOLIFE dataset. It should be noted that some context features, like traffic congestion, might be related to the day of the week. The location, velocity and traffic congestion MAEs of the context-aware methods decreased compared to the baseline method. This confirms the usefulness of location and velocity features compared to the day of the week feature. The combined context method demonstrates the best performance with the minimal MAE values for both datasets. This indicates that context information helps obtaining more accurate prediction results.

We analyze several characteristics of the movement patterns we found. The clustering results depend on segmentation threshold values and on the similarity measures. The higher they are, the more trajectories are summarized into larger MBBs. The ability to make an accurate prediction depends on the number of MBBs we have to choose from and on each MBB's size. As we can see in Table 3, context-aware methods provide smaller MBBs and more MBBs to choose from in every prediction.

Table 3. Average sizes of MBBs and candidate set

	INFATI		GEOLIFE	
	Candidates	MBB Size	Candidates	MBB Size
Baseline	3.9	1,952	4.7	1,299
Location	3.9	2,952	4.7	1,299
Velocity	9.5	1,906	7.1	1,283
Day of the week	4.5	1,721	5.4	1,352
Weather	6.1	1,853	6.2	1,465
Traffic congestion	9.8	1,622	9.3	1,002
Combined context	9.8	1,302	9.3	921

5 Conclusions

In this paper, we analyzed location prediction from a context awareness perspective. We proposed a novel way to leverage context information of several types in order to predict the future location of mobile objects. We suggested appropriate methods to extract context information based on the movement and the surrounding environment. We enhanced the k-means clustering algorithm for discovering movement patterns from spatio-temporal trajectories, to support five types of context features regarding location, velocity, day of the week, weather and traffic congestion. We designed a new algorithm for predicting the future location of a mobile object that utilizes context information based on the above ideas. The algorithm is shown to perform well, with significantly better accuracy results. Our evaluation results show that it is indeed helpful to refer to context information, while choosing the predicted movement patterns. In addition, incorporating context information allows us to collect smaller amounts of historic movement data, before we can start predicting the future location of mobile objects effectively. Thus, we can shorten the data collection period, while keeping the same level of prediction accuracy.

As future work, we suggest extending the algorithm to support other relevant context information features. For example, we can extract significant anchor points, such as home and work, from the object trajectories. We can also extract information about activities or events in the area from external data sources, like social networks. Furthermore, optimization techniques may be applied for finding the best subset of context features. In addition, as this method uses the trajectories of each mobile object alone, one may apply this technique to consider all available trajectories. Instead of using only the individual user history, we could utilize all available object trajectories together and then apply the prediction model to previously unseen users. Further work is needed for designing a formal ontology-based context model, which will be able to capture, represent, manipulate and access all characteristics of location and movement. Ontology should address such issues as semantic context representation, context dependency, and context reasoning.

References

1. Amini, S., Brush, A.J., Krumm, J., Teevan, J.: Trajectory-aware mobile search. In: Proceedings of the 2012 ACM Annual Conference on Human Factors in Computing Systems, pp. 2561–2564 (2012)
2. Kuang, L., Xia, Y., Mao, Y.: Personalized services recommendation based on context-aware QoS prediction. In: IEEE 19th International Conference in Web Services, pp. 400–406 (2012)
3. Lee, J., Han, J., Whang, K.: Trajectory clustering: a partition-and-group framework. In: Proceedings of the 2007 ACM SIGMOD International Conference on Management of Data (2007)
4. Zhang, L., Seta, N., Miyajima, H., Hayashi, H.: Fast authentication based on heuristic movement prediction for seamless handover in wireless access environment. In: IEEE WCNC, Wireless Communications and Networking Conference, pp. 2889–2893 (2007)
5. Akoush, S., Sameh, A.: Mobile user movement prediction using bayesian learning for neural networks. In: ACM IWCMC, pp. 191–196 (2007)
6. Elnekave, S., Last, M., Maimon, O.: Predicting future locations using clusters centroids. In: The Proceedings of the 15th Annual ACM International Symposium on Advances in Geographic Information Systems (2008)
7. Monreale, A., Pinelli, F., Trasarti, R., Giannotti, F.: WhereNext: a location predictor on trajectory pattern mining. In: 15th ACM SIGKDD, pp. 637–646 (2009)
8. Patel, J., Chen, Y., Chakka, V.: Stripes: an efficient index for predicted trajectories. In: SIGMOD, pp. 635–646 (2004)
9. Tao, Y., Faloutsos, C., Papadias, D., Liu, B.: Prediction and indexing of moving objects with unknown motion patterns. In: SIGMOD Conference, pp. 611–622. ACM Press (2004)
10. Jeung, H., Liu, Q., Shen, H., Zhou, X.: A hybrid prediction model for moving objects. In: IEEE 24th International Conference on Data Engineering, pp. 70–79 (2009)
11. Nizetic, I., Fertalj, K., Kalpic, D.: A Prototype for the short-term prediction of moving object's movement using Markov chains. In: Proceedings of the ITI, pp. 559–564 (2009)
12. Han, Y., Yang, J.: Clustering moving objects. In: KDD, pp. 617–622 (2004)
13. Elnekave, S., Last, M., Maimon, O.: Incremental Clustering of mobile objects. In: STDM07. IEEE (2007)
14. Schilit, B., Adams, N., Want, R.: Context-aware computing applications. In: Proceedings of the Workshop on Mobile Computing Systems and Applications. IEEE Computer Society, MD (1994)
15. Schmidt, A., Beigl, M., Gellersen, H.: There is more to context than location. In: Proceedings of the International Workshop on Interactive Applications of Mobile Computing (1998)
16. Chen, G., Kotz, D.: A survey of context-aware mobile computing research, Technical report TR2000-381, DartmouthCollege, ComputerScience (2000)

17. Sanchez, L., Lanza, J., Olsen, R., Bauer, M.: A generic context management framework for personal networking environments. In: Mobile and Ubiquitous Systems-Workshops, pp. 1–8 (2006)
18. Zhang, D., Huang, H., Lai, C.F., Liang, X.: Survey on context-awareness in ubiquitous media. Multimedia Tools Appl. **67**(1), 179–211 (2013)
19. Lopes, J., Gusmão, M., Duarte, C., Davet, P.: Toward a distributed architecture for context awareness in ubiquitous computing. J. App. Comput. Res. **3**, 19–33 (2014)
20. Zheng, Y., Zhang, L., Xie, X., Ma, W.: Mining interesting locations and travel sequences from GPS trajectories. In: Proceedings of the 18th International Conference on World Wide Web. ACM (2009)
21. Gao, H., Tang, J., Liu, H.: Mobile location prediction in spatio-temporal context. In: Nokia Mobile Data Challenge Workshop (2012)
22. Žliobaitė, I., Mazhelis, O., Pechenizkiy, M.: Context-aware personal route recognition. In: Elomaa, T., Hollmén, J., Mannila, H. (eds.) DS 2011. LNCS, vol. 6926, pp. 221–235. Springer, Heidelberg (2011)
23. Veness, C.: Calculate distance, bearing and more between points (2002). http://www.movable-type.co.uk/scripts/latlong.html
24. Jensen, C., Lahrmann, H., Pakalnis, S., Runge, J.: The INFATI Data, TimeCenter Technical report, pp. 1–10 (2004)
25. Zheng, Y., Xing, X., Ma, W.: GeoLife: A Collaborative Social Networking Service among User, location and trajectory. IEEE Data Eng. Bull. **33**(2), 32–40 (2010)

Author Index

Printed in the United States
By Bookmasters